International
REVIEW OF
Neurobiology
Volume 84

International
REVIEW OF
Neurobiology

Volume 84

Neurobiology of Dementia

EDITED BY

ALIREZA MINAGAR
Department of Neurology
Louisiana State
University Health Sciences Center
Shreveport, LA, USA

ELSEVIER

AMSTERDAM • BOSTON • HEIDELBERG • LONDON
NEW YORK • OXFORD • PARIS • SAN DIEGO
SAN FRANCISCO • SINGAPORE • SYDNEY • TOKYO
Academic Press is an imprint of Elsevier

Academic Press is an imprint of Elsevier
360 Park Avenue South, New York, NY 10010-1700
525 B Street, Suite 1900, San Diego, California 92101-4495, USA
32 Jamestown Road, London NW1 7BY, UK

This book is printed on acid-free paper.

For information on all Academic Press publications
visit our Web site at www.elsevierdirect.com

ISBN-13: 978-0-12-374833-1

Printed and bound by CPI Group (UK) Ltd, Croydon, CR0 4YY

Transferred to Digital Print 2012

CONTENTS

Neuroimaging of Cognitive Impairments in Vascular Disease

Carol Di Perri, Turi O. Dalaker, Mona K. Beyer, and Robert Zivadinov

Contributions of Neuropsychology and Neuroimaging to Understanding Clinical Subtypes of Mild Cognitive Impairment

Amy J. Jak, Katherine J. Bangen, Christina E. Wierenga, Lisa Delano-Wood, Jody Corey-Bloom, and Mark W. Bondi

Proton Magnetic Resonance Spectroscopy in Dementias and Mild Cognitive Impairment

H. Randall Griffith, Christopher C. Stewart, and Jan A. den Hollander

Application of Pet Imaging to Diagnosis of Alzheimer's Disease and Mild Cognitive Impairment

JAMES M. NOBLE AND NIKOLAOS SCARMEAS

The Molecular and Cellular Pathogenesis of Dementia of the Alzheimer's Type: An Overview

FRANCISCO A. LUQUE AND STEPHEN L. JAFFE

Alzheimer's Disease Genetics: Current Status and Future Perspectives

LARS BERTRAM

Frontotemporal Lobar Degeneration: Insights from Neuropsychology and Neuroimaging

ANDREA C. BOZOKI AND MUHAMMAD U. FAROOQ

Lewy Body Dementia

Jennifer C. Hanson and Carol F. Lippa

Dementia in Parkinson's Disease

Bradley J. Robottom and William J. Weiner

Early Onset Dementia

Halim Fadil, Aimee Borazanci, Elhachmia Ait Ben Haddou, Mohamed Yahyaoui, Elena Korniychuk, Stephen L. Jaffe, and Alireza Minagar

Normal Pressure Hydrocephalus

Glen R. Finney

Reversible Dementias

ANAHID KABASAKALIAN AND GLEN R. FINNEY

CONTRIBUTORS

Numbers in parentheses indicate the pages on which the authors' contributions begin.

Katherine J. Bangen (81), Department of Psychiatry, School of Medicine, University of California, San Diego 92093, USA

Yvonne Bannon (1), Department of Psychiatry and Behavioral Medicine, University of South Florida, 3515 East Fletcher Avenue, MDC-14, Tampa, Florida 33613, USA

Lars Bertram (167), Neuropsychiatric Genetics Group, Department of Vertebrate Genomics, Max-Planck Institute for Molecular Genetics, Berlin 14195, Germany; and Genetics and Aging Research Unit, Department of Neurology, Massachusetts General Hospital, Charlestown, Massachusetts, 02129, USA

Mona K. Beyer (49), The Norwegian Centre for Movement Disorders; and Department of Radiology, Stavanger University Hospital, Stavanger, Norway

Mark W. Bondi (81), Veterans Affairs San Diego Healthcare System, San Diego 92161, USA; and Department of Psychiatry, School of Medicine, University of California, San Diego 92093, USA

Aimee Borazanci (245), Department of Neurology, Louisiana State University School of Medicine-Shreveport, Shreveport, Louisiana 71103, USA

Andrea C. Bozoki (185), Department of Neurology, Michigan State University, East Lansing, Michigan, 48824, USA

Maria T. Caserta (1), Department of Psychiatry and Behavioral Medicine, University of South Florida, 3515 East Fletcher Avenue, MDC-14, Tampa, Florida 33613, USA

Jody Corey-Bloom (81), Department of Neurosciences, School of Medicine, University of California, San Diego 92093, USA; and Veterans Affairs San Diego Healthcare System, San Diego 92161, USA

Turi O. Dalaker (49), The Norwegian Centre for Movement Disorders, Stavanger University Hospital, Stavanger, Norway; and Department of Neurology, Buffalo Neuroimaging Analysis Center, State University of New York at Buffalo, Buffalo, New York 14203, USA

Lisa Delano-Wood (81), Veterans Affairs San Diego Healthcare System, San Diego 92161, USA; and Department of Psychiatry, School of Medicine, University of California, San Diego 92093, USA

Jack C. de la Torre (35), Sun Health Research Institute, Center for Alzheimer's Research, Sun City, Arizona 85351, USA

Carol Di Perri (49), Department of Neuroradiology, IRCCS, C. Mondino, University of Pavia, Pavia, Italy; and Department of Neurology, Buffalo Neuroimaging Analysis Center, State University of New York at Buffalo, Buffalo, New York 14203, USA

Halim Fadil (245), Department of Neurology, Louisiana State University School of Medicine-Shreveport, Shreveport, Louisiana 71103, USA

Muhammad U. Farooq (185), Department of Neurology, Michigan State University, East Lansing, Michigan, 48824, USA

Francisco Fernandez (1), Department of Psychiatry and Behavioral Medicine, University of South Florida, 3515 East Fletcher Avenue, MDC-14, Tampa, Florida 33613, USA

Glen R. Finney (263, 283), Memory and Cognitive Disorders Program, University of Florida Department of Neurology, Gainesville, Florida 32610-0236, USA

Brian Giunta (1), Department of Psychiatry and Behavioral Medicine, University of South Florida, 3515 East Fletcher Avenue, MDC-14, Tampa, Florida 33613, USA

H. Randall Griffith (105), Department of Psychology; and Alzheimer's Disease Research Center; and Department of Neurology, University of Alabama at Birmingham, Birmingham, Alabama 35233, USA

Elhachmia Ait Ben Haddou (245), Service de Neurologie B et Neurogenetique, Hopital des Specialties, Rabat, Morocco

Jennifer C. Hanson (215), Drexel University College of Medicine, Mail Stop 423, Philadelphia, Pennsylvania 19107, USA

Jan A. den Hollander (105), Department of Medicine (Cardiology), University of Alabama at Birmingham, Birmingham, Alabama 35233, USA

Stephen L. Jaffe (151, 245), Department of Neurology, Louisiana State University School of Medicine-Shreveport, Shreveport, Louisiana 71103, USA

Amy J. Jak (81), Veterans Affairs San Diego Healthcare System, San Diego 92161, USA; and Department of Psychiatry, School of Medicine, University of California, San Diego 92093, USA

Anahid Kabasakalian (283), Memory and Cognitive Disorders Program, University of Florida Department of Neurology, Gainesville, Florida 32610-0236, USA

Roger E. Kelley (21), Department of Neurology, LSU Health Sciences Center, Shreveport, Louisiana 71103, USA

Elena Korniychuk (245), Department of Neurology, Louisiana State University School of Medicine-Shreveport, Shreveport, Louisiana 71103, USA

Carol F. Lippa (215), Drexel University College of Medicine, Mail Stop 423, Philadelphia, Pennsylvania 19107, USA

Francisco A. Luque (151), VA Neurology Service, Overton Brooks VAMC, Shreveport, Louisiana 71101, USA; and Department of Neurology, Louisiana State University School of Medicine-Shreveport, Shreveport, Louisiana 71103, USA

Uma Menon (21), Department of Neurology, LSU Health Sciences Center, Shreveport, Louisiana 71103, USA

Alireza Minagar (245), Department of Neurology, Louisiana State University School of Medicine-Shreveport, Shreveport, Louisiana 71103, USA

James M. Noble (133), Gertrude H. Sergievsky Center, and the Taub Institute for Research on Alzheimer's Disease and the Aging Brain; and Department of Neurology, Columbia University Medical Center, New York 10032, USA; and Department of Neurology, Harlem Hospital Center, Columbia University College of Physicians and Surgeons, New York 10037, USA

Bradley J. Robottom (229), Department of Neurology, University of Maryland School of Medicine, Baltimore, Maryland 21230, USA

Nikolaos Scarmeas (133), Gertrude H. Sergievsky Center, and the Taub Institute for Research on Alzheimer's Disease and the Aging Brain; and Department of Neurology, Columbia University Medical Center, New York 10032, USA

Mike R. Schoenberg (1), Department of Psychiatry and Behavioral Medicine, University of South Florida, 3515 East Fletcher Avenue, MDC-14, Tampa, Florida 33613, USA

Christopher C. Stewart (105), Department of Psychology, University of Alabama at Birmingham, Birmingham, Alabama 35233, USA

Jun Tan (1), Department of Psychiatry and Behavioral Medicine, University of South Florida, 3515 East Fletcher Avenue, MDC-14, Tampa, Florida 33613, USA

William J. Weiner (229), Department of Neurology, University of Maryland School of Medicine, Baltimore, Maryland 21230, USA

Christina E. Wierenga (81), Veterans Affairs San Diego Healthcare System, San Diego 92161, USA; and Department of Psychiatry, School of Medicine, University of California, San Diego 92093, USA

Mohamed Yahyaoui (245), Service de Neurologie B et Neurogenetique, Hopital des Specialties, Rabat, Morocco

Robert Zivadinov (49), Department of Neurology, The Jacobs Neurological Institute; and Department of Neurology, Buffalo Neuroimaging Analysis Center, State University of New York at Buffalo, Buffalo, New York 14203, USA

PREFACE

The significant increase in human longevity translates into a significant increase in the number of people with dementia around the globe. Due to its relentlessly progressive course and lack of any effective treatments, neurodegenerative dementing disorders pose an enormous economic and social burden on society as well as a great emotional burden. During the last two decades, our knowledge about molecular mechanisms of the various neurodegenerative dementias has grown remarkably, giving us unprecedented insight into the fundamental mechanisms of these incurable disorders.

This volume of the *International Review of Neurobiology* focuses on dementia and related diseases. The volume contains various chapters on pathogenesis, clinical features, genetics, neuroimaging and molecular pathogenesis of the neurodegenerative dementias. The first chapter by Caserta *et al.* addresses the very important topic of normal aging and highlights its salient features. The next three chapters discuss the role of cerebrovascular disease in cognition. First, Roger E. Kelley provides a comprehensive update on subcortical ischemic cerebrovascular dementia, and then J. C. de la Torre discusses the role of cerebrovascular and cardiovascular pathology in Alzheimer's disease. These two excellent chapters are followed by another detailed chapter by Dalaker *et al.*, explaining the neuroimaging of cognitive impairment in vascular disease. These authors provide our readers with the latest developments on this subject, highlighting recent advances in neuroimaging of cognitive decline caused by cerebrovascular disorders. The focus of this volume then switches to mild cognitive impairment (MCI), and the next three contributions cover various aspects of this significant clinical issue. Jak *et al.* discuss the role of neuropsychology and neuroimaging in better understanding of clinical subtypes of MCI. The next two chapters explore various imaging procedures used to study MCI. Griffith *et al.* discuss the role of proton MR imaging in the study of dementing disorders including MCI, and then Noble and Scarmeas provide readers with an interesting perspective on the application of PET imaging for the diagnosis of Alzheimer's disease and MCI.

The next section of this volume covers Alzheimer's disease and other progressive neurodegenerative dementias. First, Luque and Jaffe discuss the underlying molecular and cellular mechanisms of dementia of the Alzheimer's type. This interesting chapter is followed by another superb update on the genetics of Alzheimer's disease by Lars Bertram. The next three chapters discuss other neurodegenerative dementias. Bozoki and Farooq present an excellent discussion concerning the neuropsychology and neuroimaging of frontotemporal lobar degeneration.

Next, Lippa and Hanson give an update on Lewy body dementia. This extensive review is followed by Robottom and Weiner's chapter on the dementia in Parkinson's disease, presenting readers with the most recent findings on this subject.

The final section of this issue of the *International Review of Neurobiology* concentrates on important subjects which we as neurologists encounter on a daily basis. First, Minagar and colleagues discuss various causes of dementia in younger individuals and present the readers with a systematic approach for the diagnosis of this uncommon presentation. Then, Glen Finney and Anahid Kabasakalian discuss the pathogenesis and clinical features, as well as the diagnostic procedures for investigating normal pressure hydrocephalus and other reversible dementias.

In summary, various contributors to this issue of the *International Review of Neurobiology* have attempted to improve the readers' knowledge concerning a broad spectrum of dementing disorders. Each contribution provides the inquisitive reader with the latest developments in our understanding of that particular topic. The editor and the very knowledgeable contributors to this issue hope that this volume of the *International Review of Neurobiology* stimulates the readers' scientific curiosity and promotes basic and clinical research in the dementia.

Finally, I would like to acknowledge the excellent contributions of the various authors to this issue of the *International Review of Neurobiology*. Without their effort and dedication, compiling this volume would have never become a reality. I would also like to thank Ms. Narmada Thangavelu and Mr. Charles Prem Kumar Neelakandan from Elsevier for their time and effort expended to process and complete on this of the journal.

ALIREZA MINAGAR

NORMAL BRAIN AGING: CLINICAL, IMMUNOLOGICAL, NEUROPSYCHOLOGICAL, AND NEUROIMAGING FEATURES

Maria T. Caserta, Yvonne Bannon, Francisco Fernandez, Brian Giunta, Mike R. Schoenberg, and Jun Tan

Department of Psychiatry and Behavioral Medicine, University of South Florida, 3515 East Fletcher Avenue, MDC-14, Tampa, Florida 33613, USA

Brain aging is characterized by numerous physiological, structural, functional, and neurocognitive changes. The interplay of these various processes is complex and characterized by large interindividual differences. Although much is not understood about how we age, there are numerous studies detailing the nature of the changes in the brain as we age. This chapter will review some of the functional, neuropsychological, neuroimaging, and immunological processes known to occur in normal or "healthy" aging. Large epidemiological studies of older adults have shown health status in the elderly is a function of the negative consequences and impairment in functional performance caused by medical co-morbidities. Immunological function declines with age, such that adaptive immunity is reduced to previously encountered pathogens, and there is a weakened adaptive immune response to novel pathogens. Structural and functional neuroimaging studies of cognitively intact older adults have consistently shown volume loss and loss of white matter structural integrity, particularly in prefrontal cortex, which may be associated with cognitive decline. While data are incomplete, one consistent finding has been a decline in cognitive domains, such as arithmetic/numerical ability and perceptual speed. Alternatively, other cognitive functions such as verbal ability, word knowledge, and semantic memory remain

INTERNATIONAL REVIEW OF
NEUROBIOLOGY, VOL. 84
DOI: 10.1016/S0074-7742(09)00401-2

quite preserved even to old age. Factors identified for healthy cognitive (brain) aging are multifactorial, and likely incorporate biological systems as well as cognitive reserve.

I. Introduction

The percentage of the population over age 65 is growing rapidly in the United States. In 2000, it was 12.4% and in 2050 it is projected to be 20.6% (Health, United States, 2005). Many people in this older age group are healthy but it is well known that functional disability increases with age. There are many theories of aging including "programmed aging" and the "wear and tear" theories. The former regards aging as being regulated by an internal clock and is seen as an extension of development, which is highly regulated and timed process. However, the interplay of disease, environment and genetics is evident not only in epidemiological studies of aging and longevity, but also in functional assessments of the elderly, and in such concepts as cognitive reserve and neural compensation.

Structural and functional imaging studies of intact older individuals have delineated specific areas where there is loss of brain volume and also a reduction in the lateralization of brain activity during cognitive performance in elderly individuals. However, not all of these changes are detrimental to maintaining performance levels and it is not yet clear which changes are precursors to further cognitive decline and which may be true compensatory mechanisms.

II. Epidemiology of Aging: Risks and Implications

Through the year 2000, approximately 35 million of people living in the USA were over 65 years. It is estimated that by 2050, this number will more than double to 82 million with 20% of the US population age 65 years or older (see Table I; Ferrucci *et al.*, 2008). With the life expectancy in old age increasing, and death rates reported to decrease, these numbers may be an underestimate of what is to come (Minino *et al.*, 2007). While older adults are generally healthy and independent and are active in various roles contributing significantly to their families and society, the increase in concurrent medical conditions causing excess morbidity and mortality as well as disability and decline in functional performance negatively impact a significant number of the elderly (Ferrucci *et al.*, 2008). Health status in the elderly is a function of the negative consequences and impairment in functional performance caused by medical co-morbidities.

TABLE I
ACTUAL AND PROJECTED GROWTH OF THE OLDER US POPULATION, 1900–2050 (MILLIONS)

		>65 Years		>85 Years	
Year	Total population (all ages)	Number	Percent of total	Number	Percent of >65 years
1900	76.1	3.1	4.1	0.1	3.2
1950	152.3	12.3	8.2	0.6	4.9
2000	276.1	34.9	12.6	4.4	12.6
2050	403.7	82.0	20.3	19.4	23.7

Note: This table from Ferrucci *et al.* (2008) is used with permission.

Any and all diseases occurring in the elderly cause excess disability and decrements in function in a variety of domains but mostly in independent function. For example, chronic health conditions such as hypertension, hyperglycemia, musculoskeletal disorders, coronary artery disease, cancer, and mixed states of anxiety and depression affect a significant number of the elderly (see Table II). By race and ethnicity these numbers vary among minority men and women when compared with that in White men and women. The proportion of these with cognitive impairment and dementing illness is also rising. Similarly, preexisting anxiety and depression may increase the risk of developing cognitive impairment (Backman, 2008; Loebach *et al.*, 2002).

In large epidemiologic studies such as the Canadian Study of Health and Aging, the authors evaluated the occurrence of an adverse event (symptom, sign, or disease), or the accumulation of a number of such co-morbid events, and modeled the events as a logistic function of chronologic age in a population (Graham *et al.*, 1999). In those who were cognitively normal, a linear relation between the log of the odds of events and chronologic age was present for the majority of symptoms and signs. Thus, the dynamics of aging are a complex process of accumulation of deficits (morbidity), whereby decline from some previously healthy level of synergistically associated symptoms and signs results in distinct patterns of disease.

But these studies are only part of the larger picture. It is a familiar epidemiologic concept that individuals vary not only in whether they develop a disease but also in the ages at which disease occurs and the rates at which pre-morbid changes progress to any specific disease. Available reports of genetic epidemiologic studies on diseases of aging are adding a new dimension—trait genomics—to assess the presence or absence of a disease (or occurrence of an adverse event) or evaluating the level of a given risk factor at one point in time (NIA Aging and Genetic Epidemiology Working Group, 2000). Moreover, two very specific types of age-specific traits are *survival traits* (the age at which a specified outcome has occurred or has not occurred) and *rate-of-change traits* (the rate at which individuals'

TABLE II
AGE-ADJUSTED[a] AND AGE-SPECIFIC MORTALITY IN THE UNITED STATES, 1950 AND 2004 AND PERCENT CHANGE

Cause of death	Age (years)	1950	2004	Change
Diseases of the heart				
Males				
	All ages	697.0	267.9	−61.6
	65–74	2292.3	723.8	−68.4
	75–84	4825.0	1893.6	−60.8
	85+	9659.8	5239.3	−45.8
Females				
	All ages	484.7	177.3	−63.4
	65–74	1419.3	388.6	−72.6
	75–84	3872.0	1245.6	−67.8
	85+	8796.1	4741.5	−46.1
Cerebrovascular disease				
Males				
	All ages	186.4	50.4	−73.0
	65–74	589.6	121.1	−79.5
	75–84	1543.6	402.9	−73.9
	85+	3048.6	1118.1	−63.3
Females				
	All ages	175.8	48.9	−72.2
	65–74	522.1	96.6	−81.5
	75–84	1462.2	374.9	−74.4
	85+	2949.4	1303.4	−55.8
Malignant neoplasms				
Males				
	All ages	208.1	227.7	9.4
	65–74	791.5	907.6	14.7
	75–84	1332.6	1662.1	24.7
	85+	1668.3	2349.5	40.8
Females				
	All ages	182.3	157.4	−13.7
	65–74	612.3	627.1	2.4
	75–84	1007.7	1023.5	1.6
	85+	1299.7	1340.1	3.1

Note: This table from Ferrucci *et al.* (2008) is used with permission.
[a]Data for "all ages" are age-adjusted using the US 2000 standard population.

characteristics change over a specified age interval). Many age-related conditions like dementing illness are likely to be highly polygenic, posing difficulties for identifying genes with these effects inclusive of the survival traits associated with longevity free of neurocognitive impairment. Moreover, there may well be a genotype that delays or accelerates onset of several diseases could affect disease expression and survival significantly. Thus, many important pathologies of aging may be completely redefined in the era of neurogenomics. Recent neuroimaging findings substantiate the neurogenomics data by way of pathologic finding in age-related changes, such as vascular stiffening and loss of muscle mass that were considered "normal" aging until recently. Genetic factors may also influence not just physiologic functions at one time point but also their rates of change with age.

Thus, with the changing health status of the elderly population, there is an ever increasing role for epidemiology. Epidemiological studies will have a major impact on not just informing the public health entities on the impact on future demand for health services but also, in the era of neurogenomics, on prevention of cognitive decline and disease states such as AD.

III. Normal Aging and Functional Performance

The biophysiological changes that occur as part of normal aging can impact and limit functional performance in the elderly. The extent of impact to functional performance in the elderly is dependant on numerous factors, including cognitive status, physical disability, and medical illness (Njegovan *et al.*, 2001), as well as, emotional factors, quality of life, and numerous protective factors (Vaillant, 2002). The concept of expressing functional performance in terms of a person's ability to carry out routine activities of daily living (ADLs) was first introduced by Katz *et al.* (1963) and expanded by Lawton and Body (1969) to include more complex functions called instrumental activities of daily living (IADLs). ADLs include basic self-care activities such as bathing, dressing, eating, transferring, ambulation, and toileting: IADLs include shopping, managing finances, cooking, cleaning, and telephone use. The US Centers for Disease Control database tracks changes in functional performance in the elderly.

The association between cognitive impairment and functional performance has been studied extensively. A number of studies have demonstrated a relationship between cognitive status and functional performance independent of social, demographic, or medical factors (Hertzog and Wallace, 1997; Moritz *et al.*, 1995; Royall *et al.*, 2005). Predicting and measuring change in ADLs and IADLs with regards to advancing age has been the focus of many studies. Katz (1965) hypothesized that older persons with progressive cognitive decline lost ability to perform ADLs in the opposite order to which the skill(s) were acquired in childhood. Additionally, functional performance of IADLs was more likely to deteriorate prior to ADLs during the course of cognitive decline associated with pathological process. Although many studies have looked at changes in functional performance, and more specifically cognitive decline in the elderly with a disease process, few studies have looked at the correlation between normal physiological cerebral aging and cognitive decline (see below).

The Freedom House Study (Royall *et al.*, 2005) compared independence in IADLs to executive functioning in 547 non-institutionalized people (community dwellers) aged 70 and older. Royall's group concluded that decline in memory had no independent association with decline in functional performance, specifically IADLs. However, decline in executive function did correlate with a subject's ability to independently perform IADLs. Other studies have observed changes in functional performance led to changes in independence, socialization, self-perception, and quality of life (Keller *et al.*, 1999).

A. Quality of Life and Protective Factors

The term "successful aging" first appeared in medical journals in 1961 (Motta *et al.*, 2005). Today the term is used to describe the absence of significant disease and disabilities, maintenance of high levels of physical and cognitive function, and the preservation of social and productive activities (Motta *et al.*, 2005). The Harvard Study of Adult Development is considered the landmark study of adult development and has provided key predictors to "successful aging." This research followed 824 individuals from three separate prospective, longitudinal studies, selected as teenagers from various facets of mental and physical health, throughout their lives. The three groups consisted of 268 Harvard sophomores, a sample of 456 Inner City youths, and 90 women from Stanford—all selected as teenagers and followed throughout their lives. The research focused on (1) adaptation to stress, mental health, and defense mechanisms; (2) the effects of habits—especially alcoholism and affective disorders upon physical health and mortality; (3) the effect of childhood risk factors upon adult adaptation; (4) the unfolding of adult development; and (5) the natural history of alcohol and substance abuse (Valliant, 2002). Although the findings are numerous, key findings suggest the following factors predictive of healthy or "successful aging": being a non-smokers; adaptive coping styles, mature defense styles, and optimism; absences of alcohol abuse; maintaining a healthy weight; stable marriage or relationships; some physical activity; social engagement; and education. Education, including attainment and childhood intellectual ability, is also thought to provide a cognitive reserve in later life (Whalley *et al.*, 2004), protecting from cognitive decline associated with normal aging, as well as, cognitive decline associated to disease process.

B. Centenarians

The number of people living beyond the age of 100 is growing throughout the industrialized world (Motta *et al.*, 2005). Several studies of centenarians have been conducted to determine their level of independence, clinical condition, and cerebral deterioration and whether centenarians are the prototype for "successful aging." The Italian Multicenter Study on Centenarians (IMUSCE) is one such study (Motta *et al.*, 2008). The IMUSCE was an epidemiological study which identified 1173 centenarians ranging in age from 100 to 109 years. A sub-sample of 346 centenarians was studied to determine the functional performance and cognitive status. The sub-sample was further divided into groups based on physical health and MMSE scores. True independence in all functional performance (IADLs) was seen in only six centenarians (1.7%): while 21 centenarians (6.1%) presented with slight dependence in functional performance. Additionally, 68 centenarians (19.67%) had only a moderate dependency in IADLs and 251 (72.6%) were severely dependent. Of the six independent functioning centenarians and the 21 centenarians with only slight dependence in IADLs, all have had significant changes in their level of social and productive activities, thus they cannot be considered prototypes of successful aging based on the current definition.

IV. T- and B-Lymphocyte Loss of Function in the Aging Immune System

With age, innate immunity progresses to a chronically active state secondary to exhaustion of the more evolutionary recent adaptive (specific) immune system (Franceschi *et al.*, 2007). This is in large part due to age-associated reduction of T-cells due to thymic involution (Aspinall 2000; Aspinall and Andrew, 2000; Linton *et al.*, 2005), as well as fewer bone marrow early progenitor B cells (Allman and Miller, 2005). In early life, naïve T-cells are activated by contact with antigens and then differentiate into effector or memory cells. Because the quantity of T-cells in healthy individuals is stable over the lifespan, peripheral T-cell turnover of preexisting populations in the thymus is required to replace cells to the system in relatively young individuals and to prime T-cells to new antigens (Fann *et al.*, 2005; Pawelec, 2005; Weng, 2006). Thus in the elderly there is an increase in the number of antigen-experienced cells and a decrease in the number of naïve T-cells in the circulation, which results in accumulation of incompetent memory lymphocytes (Cossarizza *et al.*, 1997). These cells likely clonally expand and became effector memory T-cells that were competent at one time, but then lost their antigen-specific function due to their age. This phenomenon is believed to be owed to life-long antigenic stress from immunosurveillance against persistent viruses (antigens), especially cytomegalovirus (CMV) (Solana *et al.*, 2006).

The net result of these age-associated phenomena are (1) reduced adaptive immunity to previously encountered pathogens and (2) weakened adaptive immune response to novel pathogens due to a reduction in the diversity of the antigen-recognition repertoire with age by approximately 10^8 in young adults to 10^6 in the elderly (Goronzy *et al.*, 2005). Moreover, $CD8^+$ T-cells in the elderly display significantly decreased ability to secrete interferon-gamma (IFN-γ) when stimulated by cognate antigen in comparison to younger age groups (Ouyang *et al.*, 2003a,b). Also, naïve $CD4^+$ T-cells from old humans and mice show decreased responsiveness to TCR stimulation and altered profiles of cytokine production versus naïve $CD4^+$ T-cells from young hosts. Likewise, the helper function of naïve $CD4^+$ T-cells for antibody production by B cells is also decreased (Swain *et al.*, 2005). The decline in responsiveness of naïve $CD4^+$ T-cells is due to the chronologic age of the cells themselves, and not of the individual host (Swain *et al.*, 2005), suggesting that long-term maintenance of naïve $CD4^+$ T-cells through homeostatic cytokines may not have a positive impact on their function. Indeed, naïve $CD4^+$ T-cells that have undergone homeostatic cell divisions proliferate less and secrete less IL-2 in response to antigen than do naïve $CD4^+$ T-cells that have not undergone homeostatic division as is seen in the young (Swain *et al.*, 2005).

Unlike naïve lymphocytes, memory $CD4^+$ T lymphocytes are long lived and relatively competent with age. These antigen-experienced T-cells, when isolated from healthy elderly humans and healthy old mice, respond normally to antigen-induced proliferation *in vitro* (Kovaiou *et al.*, 2005) and those generated at a young age respond well to antigens over time. Conversely, naïve $CD4^+$ T-cells derived in old age respond poorly (Haynes *et al.*, 2005). Together, these studies point to an age-associated defect in memory $CD4^+$ T-cells which may originate from defects of aged naïve $CD4^+$ T-cells that have reduced clonal diversity and proliferation potential. Interestingly, changes of the ratio of memory to effector $CD4^+$ T-cell subsets with age have also been implicated in deficient adaptive immune responses to viral infections and vaccines (Kang *et al.*, 2004).

An additional important age associated defect in T-cell function is accumulation of $CD28^-CD8^+$ T-cells and the loss of naïve $CD8^+$ T-cells. These T-cells are absent in newborns while composing some 85% of circulating $CD8^+$ T-cells in the elderly. The accumulation of $CD28^-CD8^+$ T-cells was also shown in patients with viral infections such as CMV (Almanzar *et al.*, 2005). $CD28^-CD8^+$ T-cells may signify terminal differentiation from the $CD28^+CD8^+$ subset after repeated antigenic stimulation. Functionally, these cells have a reduced proliferative response to TCR stimulation but exhibit normal or even enhanced cytotoxic capacity and are resistant to apoptosis (Azuma *et al.*, 1993). Furthermore, several studies have demonstrated the virtual disappearance of naïve $CD8^+$ T-cells in the elderly associated with a significant increase in the proportions of differentiated effector memory and effector $CD8^+$ T-cells in comparison to younger individuals. Interestingly, most of these cells in the elderly do not have short telomeres. Taken together, these data suggest that, over the course of life, these T-cell populations have undergone a process of end-stage differentiation and that persistent infection with common pathogens, such as CMV, induces chronic stimulation of specific T-cells that leads to terminal differentiation to senescent cells with an altered functional capability (Effros *et al.*, 2005; Fann *et al.*, 2005; Pawelec, 2005; Tarazona *et al.*, 2000). Further, it seems that clonal expansion of $CD28^{-}CD8^{+}$ T-cells seems directly responsible for increased infection rates and the common failed response to vaccines in the elderly (Almanzar *et al.*, 2005). Decreased adaptive immunity also involves changes in the B-cell repertoire not unlike those observed in the T-cell pool (Ghia *et al.*, 2000; Szbo *et al.*, 1999; Weksler, 2000). The quality of the humoral immune response is decreased with age owing to low serum immunoglobulin concentrations and low numbers of antigen-specific, immunoglobulin-secreting plasma cells. Further, antibody specificity, isotype and affinity changes are typical features of old age. For example, immunoglobulins produced in aged mice have lower affinity and are less protective versus those of young animals (Yang *et al.*, 1996). This may be secondary to aging's adverse effects on the germinal centre reaction in secondary lymphoid tissues (Zheng *et al.*, 1997) and leads to a diminished quantity of germinal centers in response to tetanus toxoid stimulation (Kraft *et al.*, 1987). The quantity of B cells also decreases in the elderly (Franchesci *et al.*, 1995). At the cellular level, alteration in immunoglobulin generation (through class switching) in B cells is observed in aged individuals (Frasca *et al.*, 2005), which may contribute to the decline of the adaptive immune response in the elderly.

V. Neuropsychology of Aging

The literature is replete with data indicating a broad spectrum of neuropsychological functions decline with normal aging (e.g., Craik and Salthouse, 2000; Woodruff-Pak, 1997). What is less clear is the line separating normal aging from cognitive dysfunction and dementia (e.g., Morris and Cummings, 2005; Morris *et al.*, 2001; Petersen, 2003; Raz, 1996, 2000). Various terms and criterion have been used to separate normal aging from abnormal, such as age associated memory impairment (AAMI; Crook *et al.*, 1986), aging-associated cognitive decline (Levy, 1994), benign senescent forgetfulness (Kral, 1962), cognitive impairment no dementia (CIND; Graham *et al.*, 1997), mild cognitive disorder (World Health Organization, 1978), mild cognitive impairment (MCI; Morris *et al.*, 2001; Petersen *et al.*, 1999), mild neurocognitive disorder (Gutierrez *et al.*, 1993), and questionable dementia (Hughes *et al.*, 1982), among others (see also Craik and Salthouse, 2000 for a review). Further complicating the division between normal and abnormal aging processes is the variability in neuropathological and neuropsychological functions between and within samples (e.g., see Craik and Salthouse, 2000; Woodruff-Pak, 1997 for reviews). A summary of neuropsychological changes associated with aging is provided below.

A. Changes in Neuropsychological Function in Normal Aging

It has long been recognized that some cognitive processes decline with age while others appear less affected, giving rise to the traditional dichotomy between the so-called "crystallized" abilities (e.g., accumulated knowledge and expertise) versus those termed "fluid" abilities (e.g., fluid reasoning, working memory, visuoperceptual abilities, processing speed, etc.) (e.g., Craik and Salthouse, 2000; Woodruff-Pak, 1997 for review). Within this framework, crystallized abilities have been considered generally stable with increasing age, while fluid abilities decline with increasing age, some thought to begin declining as early as in one's 20s (Babcock, 1930; Craik and Salthouse, 2000; Lezak, 1995; Wechsler, 1958; Woodruff-Pak, 1997 for review). Until recently, however, much of the previous knowledge regarding age-related cognitive decline have been criticized for lack of longitudinal data, with most derived from cross-sectional studies or other studies having small samples and/or short follow-up time frames (e.g., Baltes and Mayer, 1999; Hertzog *et al.*, 2003; Schaie, 1996).

Changes in normal aging are now being understood from longitudinal data from several sources, including the Berlin Aging Study (Baltes and Mayer, 1999), Mayo's Older Americans Normative Studies (MOANS; Ivnik *et al.*, 1992), Seattle Longitudinal Study (Schaie, 1996), and Victoria Longitudinal Study (Dixon and de Frias, 2004; Hertzog *et al.*, 2003). Combined, these studies provide longitudinal data of individuals beginning in participants' 20s up to 100 years old and beyond (e.g., Singer *et al.*, 2003). While data are incomplete, one consistent finding has been a decline in select cognitive domains beginning as early as the middle to late 20s. Alternatively, other cognitive functions remain quite preserved even to very old age (Singer *et al.*, 2003). Schaie (1994) found the average performances of arithmetic/numerical ability and perceptual speed beginning to decline at around age 25 years old. Alternatively, a general decline in performance on tasks of inductive reasoning, verbal ability, and episodic memory did not become apparent until the fifth or sixth decade of life. While episodic memory is known to decrease with age, some research suggesting as early as the third or fourth decade of life, other research suggests the decline may be much less precipitous and is gradual, at least until the seventh decade of life (Backman *et al.*, 2000). Similarly, Singer *et al.* (2003) found performance on tasks of processing speed, episodic memory, and verbal fluency declined from 70 to 100 years old. Alternatively, word knowledge has consistently demonstrated resilience to aging, remaining quite stable from one's 70s through 100+ years old (e.g., Lezak, 1995; Singer *et al.*, 2003). Executive functions (rapid problem solving, verbal fluency, inhibition, and flexibility) show early and considerable age-related declines (Mittenberg *et al.*, 1989; Parkin, 1996). Indeed, Mittenberg *et al.* (1989) found performance on neuropsychological tests of executive skills, thought to measure frontal lobe functioning, exhibited more age-related decline than other tests in a comprehensive battery. Further complicating understanding age-related decline in cognitive function are data establishing individual variability in cognitive performance can vary considerably over 1–3 year time periods, although group level change tends to be more modest, at least to the sixth or seventh decades of life (e.g., Dixon *et al.*, 2004; Ivnik *et al.*, 1999). Recent data suggest increased individual variability in cognitive function predicts mortality (MacDonald *et al.*, 2008).

In summary, the cognitive decline observed in normal aging reflects the complex brain changes that occur with increasing age. The decline of cognitive abilities is a complex tapestry of selective areas of preservation while other interrelated areas decline all against a backdrop of biological and physical changes. Indeed, considerable research suggests two over-arching cognitive abilities that decrease with age, processing speed, and working memory, can account for a majority of variance in the age-related declines across a variety of cognitive skills (e.g., see Meyerson *et al.*, 1990; Mitchell, 1993; Salthouse, 1996). As an example, Hertzog *et al.* (2003) found deficits in episodic memory could be accounted for by declines in working memory and perceptual speed. Thus, declines in episodic memory, numerical ability/arithmetic, verbal fluency, and problem solving are, themselves, moderated by the decline in processing speed and working memory. It has been proposed that the decline in processing speed may be associated with the structural integrity of cerebral white matter, but rather than being linear, reflects a threshold phenomenon, in which age-related declines after white matter dysfunction exceeds a critical level (Boone *et al.*, 1992; DeCarli *et al.*, 1995; Raz, 2000).

B. Staving off Cognitive Decline

Aerobic exercise has shown to decrease cognitive morbidity in aging, but a more comprehensive picture including cognitive reserve and cognitive engagement is likely accounting for preventing cognitive morbidity (e.g., Milgram *et al.*, 2006; Stern, 2006). Research has also found older adults routinely use informal techniques to improve memory (e.g., increase use of external memory aides and organizational skills) in compensating for daily cognitive challenges (Dixon *et al.*, 2001). See discussion relating cognitive reserve to structural and functional neuroanatomy below.

VI. Imaging Studies in Aging

The changes that occur with aging in the brain are complex and are associated with large interindividual variability. There is, however, a pattern of selective loss and preservation delineated in numerous studies. The structural and functional deterioration in specific brain areas is thought to be responsible for the reported cognitive decline that occurs with age while the substrates of "cognitive reserve" are less well defined.

Longitudinal studies using structural MRI from age 3 to 70 years old have shown an increase in cortical gray matter until, about, age 5 followed by a gradual decline in volume until age 70 (Pfefferbaum *et al.*, 1994). White matter growth accelerates during adolescence and then plateaus in the third decade. Several studies have indicated significant gray matter tissue volume loss (Blatter *et al.*, 1995; Brickman *et al.*, 2005; Sullivan *et al.*, 2004; Thompson *et al.*, 2003) while others have shown greater white matter loss in aging (Guttmann *et al.*, 1998; Jernigan *et al.*, 2001).

In general, brain atrophy has been reported consistently in aging and it is more pronounced in the frontal brain areas (DeCarli *et al.*, 1994; Raz *et al.*, 1997; Resnick *et al.*, 2003; Saloat *et al.*, 1999; Tisserand *et al.*, 2002). Other brain areas affected by aging are the temporal lobes with relative sparing of the parietal and occipital lobes (Bartzokis *et al.*, 2001; Good *et al.*, 2001; Raz *et al.*, 1997, 2005). The question of whether this loss is primarily due to gray matter or white matter volume loss remains controversial (see Sullivan and Pfefferbaum, 2007).

Although controversial, hippocampal volume is thought to remain relatively stable with age (Shamy *et al.*, 2006; Sullivan *et al.*, 1995, 2001, 2005) and remains intact in size in healthy elderly adults without concurrent hypertension (Raz *et al.*, 2005). There is age-related dilation of the temporal horns and this has been interpreted as hippocampal volume decline. It appears that volume loss and CSF expansion occur in diseases such as Alzheimer's, but not in normal aging. The relative resilience of hippocampal volume to aging makes hippocampal deviations from normal size a sensitive indicator of pathology, especially of the Alzheimer type.

Studies of normal aging have benefited immeasurably by the introduction of quantitative DTI, which has successfully revealed evidence of microstructural disruption of regional white matter even in regions appearing normal on volume imaging. MR diffusion-weighted imaging (DWI) and DTI allow quantification of microscopic water movement. In regions with few or no physical boundaries, such as CSF in the ventricles, water movement is random, that is, freely diffusing, and is therefore isotropic. By contrast, the path of a water molecule in white matter is constrained by physical boundaries, such as the axon sheath, causing the movement to be greater along the long axis of the fiber than across it and is anisotropic, typically measured as fractional anisotropy (FA) and ranging between 0 and 1 on a normalized scale (Pierpaoli and Basser, 1996). DTI is therefore sensitive to the detection of tightly packed fibers in locally parallel orientation, characterizing white matter tracts in the brain. One of the most robust findings regarding age-related differences in FA has been the finding of low FA in frontal white matter (Sullivan *et al.*, 2001). This finding has been supported by postmortem investigations and animal models of aging (for review see Sullivan and Pfefferbaum, 2007) and suggests that decreased brain white matter in older individuals may be a consequence of myelin sheath changes and/or accumulation of fluid in fiber tracts. These types of changes may be reflected as increased white matter hyperintensities observed on T2-weighted MRI images in over 90% of older adults.

Beyond the structural changes summarized so far, there have been numerous functional imaging studies that have described cognitive changes with aging. These studies using positron emission tomography, PET, and functional magnetic resonance imaging, fMRI, have revealed how the neural correlates of different cognitive functions change as a function of aging. Most studies show a reduction in the lateralization of brain activity during cognitive performance in elderly individuals. For example, in a study utilizing word pair encoding, older adults showed weaker activity in several left prefrontal cortical (PFC) areas while several PFC regions showed increased activity (Anderson *et al.*, 2000). During episodic memory retrieval, age-related decreases were typically found in the right PFC and right parietal regions, while age-related increases in activation were found in left PFC as well as bilateral anterior cingulate and cuneus/precuneas regions (Anderson *et al.*, 2000; Cabeza *et al.*, 1997, 2002). Some investigators postulated that the age-related asymmetry reductions found in multiple studies reflect a compensatory mechanism. Older adults engage both hemispheres for tasks that younger adults usually use one hemisphere for, to compensate for neurocognitive deficits.

In a longitudinal study of 25 healthy older adults assessing changes in PET rCBF cerebral blood flow over an 8 year period among healthy older adults found no significant declines in verbal and figural recognition memory accuracy over an 8 year period (Beason-Held *et al.*, 2008). There were regional increases and decreases in brain activity were found over time in these individuals, particularly PFC regions. These data support the cross-sectional studies described previously where altered patterns of cerebral activity were found in old relative to young individuals and extend these findings to show that changing brain function continues throughout later life in cognitively stable individuals.

Although it is clear older adults demonstrate deficits relative to younger adults in various aspects of learning and memory by neuroimaging correlates, several studies have suggested there is differential susceptibility to age-related changes and dementia related to variables like education, IQ, and engagement in leisure activities (Gold *et al.*, 1995; Hultcsch *et al.*, 1999; Scarmeas *et al.*, 2001; Stern *et al.*, 1994). These studies provide epidemiological evidence for the presence of cognitive reserve (Stern, 2006). Cognitive reserve suggests individual differences in how tasks are processed might provide some protection against age-related changes or progressing brain pathology. Our understanding of how cognitive reserve might be reflected by brain networks is ongoing. Stern and his colleagues (2005) identified network activation changes which involve the right hippocampus, posterior insula, and right and left operculum as well as the inferior parietal lobe and association cortex, left posterior cingulated, and right and left calcarine cortex. Young individuals activation patterns in these networks were different from older individuals performing the same task at the same level (Stern *et al.*, 2005). These authors suggest that cognitive reserve consists of two separate components: neural reserve and neural compensation. Neural reserve may result from innate differences and/or be modulated through life events such as educational experience. The substrate for this may be neural networks that are highly efficient or have greater capacity in the face of higher demand. Neural compensation on the other hand is defined as a change in neural network use such that different brain networks are used in a specific task, due to the physiological effects of aging or brain pathology. The study of the neural substrates of cognitive reserve is in its infancy, but imaging techniques assessing both functional and structural changes in the brains of older adults are progressing rapidly.

References

Allman, D., and Miller, J. P. (2005). B cell development and receptor diversity during aging. *Curr. Opin. Immunol.* **17,** 463–467.

Almanzar, G., Schwaiger, S., Jenewein, B., Keller, M., Herndler-Brandstetter, D., Wurzner, R., Schonitzer, D., Grubeck-Lobenstein, B., and Almanzar, G. (2005). Long-term cytomegalovirus infection leads to significant changes in the composition of the CD8+ T-cell repertoire, which may be the basis for an imbalance in the cytokine production profile in elderly persons. *J. Virol.* **79,** 3675–3683.

American Psychiatric Association. (2000). "Diagnostic and Statistical Manual of Mental Disorders," 4th Ed. text revision ed. American Psychiatric Association, Washington, DC.

Anderson, N. D., Iidaka, T., McIntosh, A. R., Kapur, S., Cabeza, R., and Craik, F. I. M. (2000). The effects of divided attention on encoding- and retrieval related brain activity. A PET study of younger and older adults. *J. Cogn. Neurosci.* **12,** 775–792.

Aspinall, R. (2000). Longevity and the immune response. *Biogerontology* **1,** 273–278.

Aspinall, R., and Andrew, D. (2000). Immunosenescence: Potential causes and strategies for reversal. *Biochem. Soc. Trans.* **28,** 250–254.

Azuma, M., Phillips, J. H., and Lanier, L. L. (1993). CD28- T lymphocytes antigenic and functional properties. *J. Immunol.* **150,** 1147–1159.

Babcock, H. (1930). An experiment in the measurement of mental deterioration. *Arch. Psych.* **18,** 5–105.

Backamn, L., Small, B. J., Wahlin, A., and Larsson, M. (2000). Cognitive functioning in very old age. *In* "The handbook of aging and cognition" (F. I. M. Craik and T. A. Salthouse, Eds.), 2nd, pp. 499–558. Laurence Erlbaum Associates, Hillsdale, NJ.

Backman, L. (2008). Memory and cognition in preclinical dementia: What we know and what we do not know. *Can. J. Psych.* **53**(6), 354–360.

Baltes, P. B., and Mayer, K. U. (1999). "The Berlin. Aging Study: Aging from 70 to 100." Cambridge University Press, Cambridge.

Bartzokis, G., Beckson, M., Lu, P. H., Neuchertlein, K. H., Edwards, N., and Mintz, J. (2001). Age related changes in temporal lobe volumes in men: A magnetic resonance imaging study. *Arch. Gen. Psychiatry* **58,** 461–465.

Beason-Held, L. L., Kraut, M. A., and Resnick, S. M. (2008). Longitudinal changes in aging brain function. *Neurobiol. Aging* **29,** 483–496.

Blatter, D. D., Bigler, E. D., Gale, S. D., *et al.* (1995). Quantitative volumetric analysis of brain MR: Normative database spanning 5 decades of life. *AJNR Am. J. Neuroradiol.* **16,** 241–251.

Brickman, A. M., Buchsbaum, M. S., Shihabuddin, L., *et al.* (2005). Age associated change in orbital, cingulate, and dorsolateral frontal lobe gray and white matter volume [abstract]. *J. Int. Neuropsychol. Soc.* **11,** 176.

Boone, K. B., Miller, B. L., Lesser, I. M., Mehringer, C. M., Hill-Gutierrez, E., Goldberg, M. A., and Berman, N. G. (1992). Neuropsychological correlates of white-matter lesions in healthy elderly subjects. A threshold effect. *Arch. Neurol.* **49,** 549–554.

Cabeza, R., Grady, C. L., Nyberg, L., McIntosh, A. R., Tulving, E., Kapus, S., Jennings, J. M., Houle, S., and Craik, F. I. M. (1997). Age-related differences in neural activity during memory encoding and retrieval. A positron emission tomography study. *J. Neurosci.* **17,** 391–400.

Cabeza, R., Anderson, N. D., Locantore, J. K., and McIntosh, A. R. (2002). Aging gracefully: Compensatory brain activity in high-performing older adults. *NeuroImage* **17,** 1394–1402.

Cossarizza, A., Ortolani, C., Paganelli, R., Barbieri, D., Monti, D., Sansoni, P., Fagiolo, U., Castellani, G., Bersani, F., Londei, M., Franceschi, C., Cossarizza, A., *et al.* (1996), CD45 isoforms expression on CD4+ and CD8+ T cells throughout life, from newborns to centenarians: Implications for T cell memory. *Mech. Ageing Dev.* **86,** 173–195.

Craik, F. I. M., and Salthouse, T. A. (2000). "The Handbook of Aging and Cognition," 2nd Ed. Lawrence Erlbaum Associates Inc. Publishers, Mahwah, NJ.

Crook, T., Bartus, R. T., Ferris, S. H., Whitehouse, P., Cohen, G. D., and Gershon, S. (1986). Age-associated memory impairment: Proposed diagnostic criteria and measures of clinical change. Report of a National Institute of Mental Health. Work. Group. *Dev. Neuropsychol.* **2,** 261–276.

DeCarli, C., Murphy, D. G., Gillette, J. A., *et al.* (1994). Lack of age-related differences in temporal lobe volume of very healthy adults. *AJNR Am. J. Neuroradiol.* **15,** 689–696.

DeCarli, C., Murphy, D. G. M., Tran, M., Grady, C. L., Haxby, J. V., Gillette, J. A., Salerno, J. A., Gonzales-Aviles, A., Horwitz, B., Rapoport, S. I., and Schapiro, M. B. (1995). The effect of white matter hyperintensity volume on brain structure, cognitive performance and cerebral metabolism in 51 healthy adults. *Neurology* **45,** 2077–2084.

Dixon, R. A., and de Frias, C. M. (2004). The Victoria Longitudinal Study: From characterizing cognitive aging to illustrating changes in memory compensation. *Aging Neuropsych. Cogn.* **11,** 346–376.

Dixon, R. A., de Frias, C. M., and Bäckman, L. (2001). Characteristics of self-reported memory compensation in late life. *J. Clin. Exp. Neuropsychol.* **23,** 650–661.

Dixon, R. A., Wahlin, A., Maitland, S. B., Hultsch, D. F., Hertzog, C., and Backman, L. (2004). Episodic memory change in late adulthood: Generalizability across samples and performance. *Mem. Cogn.* **32**(5), 768–778.

Effros, R. B., Dagarag, M., Spaulding, C., and Man, J. (2005). The role of $CD8^+$ T-cell replicative senescence in human aging. *Immunol. Rev.* **205,** 147–157.

Ferrucci, L., Giallauria, F., and Guralnik, J. M. (2008). Epidemiology of aging. *Radiol. Clin. N. Am.* **46,** 643–652.

Franceschi, C., Capri, M., Monti, D., Giunta, S., Olivieri, F., Sevini, F., Panourgia, M. P., Invidia, L., Celani, L., Scurti, M., Cevenini, E., Castellani, G. C., *et al.* (2007). Inflammaging and anti-inflammaging: A systemic perspective on aging and longevity emerged from studies in humans. *Mech. Ageing Dev.* **128,** 92–105.

Franceschi, C., and Cossarizza, A. (1995). Introduction: The reshaping of the immune system with age. *Int. Rev. Immunol.* **12,** 1–4.

Frasca, D., Riley, R. L., and Blomberg, B. B. (2005). Humoral immune response and B-cell functions including immunoglobulin class switch are downregulated in aged mice and humans. *Semin. Immunol.* **17,** 378–384.

Ghia, P., Melchers, F., and Rolink, A. G. (2000). Age-dependent changes in B lymphocyte development in man and mouse. *Exp. Gerontol.* **35,** 159–165.

Gold, D. P., Andreas, D., Etezadi, J., Arbuckle, T., Schwartman, A., and Chailelson, J. (1995). Structural equation model of intellectual change and continuity and predictors of intelligence in older men. *Psychol. Aging* **10,** 294–303.

Goronzy, J. J., and Weyand, C. M. (2005). T cell development and receptor diversity during aging. *Curr. Opin. Immunol.* **17,** 468–475.

Graham, J. E., Rockwood, K., Beattie, B. L., Eastwood, R., Gauthier, S., Tuokko, H., *et al.* (1997). Prevalence and severity of cognitive impairment with and without dementia in an elderly population. *Lancet* **349**(9068), 1793–1796.

Graham, J. E., Mitnitski, A. B., Mogilner, A. J., and Rockwood, K. (1999). Dynamics of cognitive aging: Distinguishing functional age and disease from chronologic age in a population. *J. Epidemiol.* **150,** 1045–1054.

Gutierrez, R., Atkinson, J. H., and Grant, I. (1993). Mild neurocognitive disorder: Needed addition to the nosology of cognitive impairment (organic mental) disorders. *J. Neuropsychiatry Clin. Neurosci.* **5,** 161–177.

Guttmann, C. R., Jolesz, F. A., Kikinis, R., *et al.* (1998). White matter changes with normal aging. *Neurology* **50,** 972–978.

Haynes, L., Eaton, S. M., Burns, E. M., Randall, T. D., and Swain, S. L. (2005). Newly generated CD4 T cells in aged animals do not exhibit age-related defects in response to antigen. *J. Exp. Med.* **201,** 845–851.

Hertzog, D. J., and Wallace, R. B. (1997). Measures of cognitive functioning in the AHEAD study. *J. Gerontol. Psychol. Sci. Soc. Sci.* **52B,** 37–48.

Hertzog, C., Dixon, R. A., Hultsch, D. F., and MacDonald, S. W. S. (2003). Latent change models of adult cognition: Are changes in processing speed and working memory associated with changes in episodic memory? *Psychol. Aging* **18,** 755–770.

Hughes, C. P., Berg, L., *et al.* (1982). A new clinical scale for the staging of dementia. *Brit. J. Psychiatry* **140,** 566–572.

Hultsch, D. F., Hertzog, C., Small, B. J., and Dixon, R. A. (1999). Use it or lose it? Engaged life style as a buffer of cognitive decline in aging. *Psychol. Aging 19* **14**(2), 245–263.

Ivnik, R. J., Malec, J. F., Smith, G. E., Tangalos, E. G., Petersen, R. C., Kokmen, E., *et al.* (1992). Mayo's older Americans normative studies: WAIS-R norms for ages 56 to 97. *Clin. Neuropsychol.* **6,** 1–30.

Ivnik, R. J., Smith, G. E., Lucas, J. A., Petersen, R. C., Boeve, B. F., Kokmen, E., *et al.* (1999). Testing normal older people three or four times at 1- to 2-year intervals: Defining normal variance. *Neuropsychology* **13**(1), 121–127.

Jernigan, T. L., Archibald, S. L., Fenema-Notestine, C., Gamst, A. C., Stout, J. C., Bonner, J., and Hesselink, J. R. (2001). Effect of age on tissues and regions of cerebrum and cerebellum. *Neurobiol. Aging* **22,** 581–594.

Kang, I., Hong, M. S., Nolasco, H., Park, S. H., Dan, J. M., Choi, J. Y., and Craft Kang. (2004). I: Age-associated change in the frequency of memory CD4+ T cells impairs long term CD4+ T cell responses to influenza vaccine. *J. Immunol.* **173,** 673–681.

Katz, S., Ford, A. B., Moskowitz, R. W., Jackson, B. A., and Jaffe, M. W. (1963). The index of ADL: A standardized measure of biological and psychosocial function. *JAMA* **185,** 914–919.

Keller, B. K., Morton, J. L., Thomas, V. S., and Potter, J. F. (1999). The effect of visual and hearing impairments on functional status. *J. Am. Geriatr. Soc.* **47,** 1319–1325.

Kovaiou, R. D., Weiskirchner, I., Keller, M., Pfister, G., Cioca, D. P., and Grubeck-Loebensten, B. (2005). Age-related differences in phenotype and function of CD4+ T cells are due to a phenotypic shift from naive to memory effector CD4+ T cells. *Int. Immunol.* **17,** 1359–1366.

Kraft, R., Bachmann, M., Bachmann, K., Buerki, H., Hess, M. W., Cottier, H., and Stoner, R. D. (1987). Satisfactory primary tetanus antitoxin responses but markedly reduced germinal centre formation in first draining lymph nodes of ageing mice. *Clin. Exp. Immunol.* 447–453.

Kral, V. A. (1962). Senescent forgetfulness: Benign and malignant. *Can. Med. Assoc. J.* **86,** 257–260.

Lawton, M. P., and Body, E. M. (1969). Assessment of older people: Self-maintaining and instrumental activities of daily living. *Gerontologist* **9,** 179–186.

Levy, R. (1994). Aging-associated cognitive decline. *Int. Psychogeriatr.* **6,** 63–68.

Lezak, M. (1995). Neuropsychological Assessment Third Edition. Oxford University Press, New York.

Linton, P. J., Li, S. P., Zhang, Y., Bautista, B., Huynh, Q., Trinh, T., and Linton, P. J. (2005). Intrinsic versus environmental influences on T-cell responses in aging. *Immunol. Rev.* **205,** 207–219.

Loebach Wetherell, J., Reynolds, C. A., Gatz, M., and Pederson, N. L. (2002). Anxiety, cognitive performance, and cognitive decline in normal aging. *J. Gerontol. Psychol. Sci.* **57B**(3), 246–255.

MacDonald, S. W. S., Hultsch, D. F., and Dixon, R. A. (2008). Predicting impending death: Inconsistency in speed is a selective and early marker. *Psychol. Aging* **23,** 595–607.

Milgram, N. W., Siwak-Tapp, C. T., Araujo, J., and Head, E. (2006). Neuroprotective effects of cognitive enrichment. *Ageing Res. Rev.* **5,** 354–369.

Minino, A. M., Heron, M., Smith, B. L., *et al.* (2007). Deaths: Final data for 2004. *Nat. Vital Stat. Rep.* **55**(9), 1–120.

Mittenberg, W., Seidenberg, M., O'Leary, D. S., and DiGiulio, D. V. (1989). Changes in cerebral functioning associated with normal aging. *J. Clin. Exp. Neuropsychol.* **11,** 918–932.

Mitchell, D. B. (1993). Implicit and explicit memory for pictures: Multiple views across the lifespan. *In* "Implicit Memory: New directions in Cognition, development, and neuropsychology" (P. Graf and M. E. J. Mason, Eds.), pp. 171–190. Laurence Erlbaum Associates, Hillsdale, NJ.

Morris, J. C., and Cummings, J. (2005). Mild cognitive impairment (MCI) represents early-stage Alzheimer's disease. *J. Alzheimers Dis.* **7**(3), 235–239; discussion 255-262.

Morris, J. C., Storandt, M., Miller, J. P., McKeel, D. W., Price, J. L., Rubin, E. H., *et al.* (2001). Mild cognitive impairment represents early-stage Alzheimer disease. *Arch. Neurol.* **58**(3), 397–405.

Mortiz, D. J., Kasl, S. V., and Berkman, L. F. (1995). Cognitive functioning and the incidence of limitations in activities of daily living in an elderly community sample. *Am. J. Epidemiol.* **141,** 41–49.

Motta, M., Bennati, E., Ferlito, L., Malaguarnera, M., and Motta, L. (2005). Successful aging in centenarians: Myths and reality. *Arch. Gerontol. Geriat.* **40,** 241–251.

Motta, M., Ferlito, L., Magnolfi, S. U., Petruzzi, E., Pinzani, P., Malentacchi, F., Petruzzi, I., Bennati, E., and Malaguarnera, M. (2008). Cognitive and functional status in the extreme longevity. *Arch. Gerontol. Geriat.* **46,** 245–252.

Myerson, J., Hale, S., Wagstaff, D., Poon, L. W., and Smith, G. A. (1990). The information-loss model: A mathematical theory of age-related cognitive slowing. *Psychol. Rev.* **97,** pp. 475–487.

National Institute on Aging and Genetic Epidemiology Working Group. (2000). http://www.nih.gov/nia/conferences/GeneticReport111199.htm.

Njegovan, V., Man-Son-Hing, M., Mitchell, S. L., and Molnar, F. J. (2001). The hierarchy of functional loss associated with cognitive decline in older persons. *J. Gerontol.* **56A**(10), M638–M643.

Ouyang, Q., Wagner, W. M., Voehringer, D., Wikby, A., Klatt, T., Walter, S., Muller, C. A., Pircher, H., and Pawelec, G. (2003a). Age-associated accumulation of CMV-specific CD8+ T cells expressing the inhibitory killer cell lectin-like receptor G1 (KLRG1). *Exp. Gerontol.* **38,** 911–920.

Ouyang, Q., Wanger, W. M., Wikby, A., Walter, S., Aubert, G., Dodi, A. I., Travers, P., Pawelec, G., and Ouyan, Q. (2003b). Large numbers of dysfunctional CD8+ T lymphocytes bearing receptors for a single dominant CMV epitope in the very old. *J. Clin. Immunol.* **23,** 247–257.

Pawelec, G. (2005). When T cells get old. *Sci. Aging Knowl. Environ.* **2005,** 39.

Parkin, A. J. (1996). Explorations in cognitive neuropsychology. Blackwell Publishers, Oxford.

Petersen, R. C. (2003). "Mild Cognitive Impairment". Oxford Press, New York.

Petersen, R. C., Smith, G. E., Waring, S. C., Ivnik, R. J., Tangalos, E. G., and Kokmen, E. (1999). Mild cognitive impairment: Clinical characterization and outcome. *Arch. Neurol.* **56**(3), 303–308.

Pfefferbaum, A., Mathalon, D. H., Sullivan, E. V., Rawles, J. M., Zipursky, R. B., and Lim, K. O. (1994). A quantitative magnetic resonance imaging study of changes in brain morphology from infancy to late adulthood. *Arch. Neurol.* **51,** 874–887.

Pierpaoli, C., and Basser, P. J. (1996). Towards a quantitative assessment of diffusion anisotropy. *Magn. Reson. Med.* **36,** 893–906.

Raz, N. (1996). Neuroanatomy of aging brain: Evidence from structural MRI. *In* "Neuroimaging II: Clinical Applications" (E. D. Bigler, Ed.), pp. 153–182. Academic Press, New York.

Raz, N. (2000). Aging of the brain and it's impact on cognitive performance: Integration of structural and functional findings. *In* "The Handbook of Aging and Cognition" (F. I. M. Craik and T. A. Salthouse, Eds.), 2nd, pp. 1–90. Lawrence Erlbaum Associates Inc. Publishers, Mahwah, NJ.

Raz, N., Gunning, F. M., Head, D., *et al.* (1997). Selective aging of the human cerebral cortex observed *in vivo*: Differential vulnerability of the prefrontal gray matter. *Cereb. Cortex* **7,** 268–282.

Raz, N., Lindenberger, U., Rodrigue, K. M., Kennedy, K. M., Head, D., and Williamson, A. (2005). Regional brain changes in aging healthy adults: General trends, individual differences, and modifiers. *Cereb. Cortex* **15,** 1676–1689.

Resnick, S. M., Pham, D. L., Kraut, M. A., *et al.* (2003). Longitudinal magnetic resonance imaging studies of older adults: A shrinking brain. *J. Neurosci.* **23,** 3295–3301.

Royall, D. R., Palmer, R., Chiodo, L. K., and Polk, M. J. (2005). Normal rates of cognitive change in successful aging: The freedom house study. *J. Int. Neuropsychol. Soc.* **11,** 899–909.

Salthouse, T. A. (1996). The processing speed theory of adult age differences in cognition. *Psychol. Rev.* **103,** 403–428.

Scarmeas, N., Levy, G., Tang, M. X., Manly, J., and Stern, Y. (2001). Influence of leisure activity on the incidence of Alzheimer's disease. *Neurology* **12,** 2236–2242.

Schaie, K. W. (1994). The course of adult intellectual development. *Am. Psychol.* **49,** 304–313.

Schaie, K. W. (1996). Intellectual development in adulthood: The Seattle longitudinal study. Cambridge University Press, New York.

Shamy, J. L., Buonocore, M. H., Makaron, L. M., Amaral, D. G., Barnes, C. A., and Rapp, P. R. (2006). Hippocampal volume is preserved and fails to predict recognition memory in aged rhesus monkeys (*Macaca mulatta*). *Neurobiol. Aging* **27,** 1405–1415.

Singer, T., Verhaeghen, P., Ghisletta, P., Lindenberger, U., and Baltes, P. B. (2003). The fate of cognition in very old age: Six-year longitudinal findings in the Berlin Aging Study [BASE]. *Psychol. Aging* **18,** 318–331.

Stern, Y. (2006). Cognitive reserve and Alzheimer disease. *Alzheimer Dis. Assoc. Disord.* **20,** 112–117.

Stern, Y., Harbeck, C., Moeller, J., Scarmeas, N., Anderson, K. E., Hiltron, H. J., Flynn, J., Sackheim, H., and van Heertum, R. (2005). Brain networks associated with cognitive reserve in healthy young and old adults. *Cerebr. Cortex* **15,** 394–402.

Sullivan, E. V., and Pfefferbaum, A. (2007). Neuroradiological characterization of normal adult aging. *Brit. J. Radiol.* S99–S108.

Sullivan, E. V., Marsh, L., Mathalon, D. H., Lim, K. O., and Pfefferbaum, A. (1995). Age-related decline in MRI volumes of temporal lobe gray matter but not hippocampus. *Neurobiol. Aging* **16,** 591–606.

Sullivan, E. V., Pfefferbaum, A., Swan, G. E., and Carmelli, D. (2001). Heritability pf hippocampal size in elderly twin men, equivalent influence from genes and environment. *Hippocampus* **11,** 754–762.

Sullivan, E. V., Rosenbloom, M., Serventi, K. L., and Pfefferbaum, A. (2004). Effects of age and sex on volumes of the thalamus, pons, and cortex. *Neurobiol. Aging* **25,** 185–192.

Sullivan, E. V., Marsh, A. L., and Pfefferbaum, A. L. (2005). Preservation of hippocampal volume through adulthood in healthy men and women. *Neurobiol. Aging* **26,** 1093–1098.

Swain, S., Clise-Dwyer, K., and Haynes, L. (2005). Homeostasis and the age-associated defect of CD4 T cells. *Semin. Immunol.* **17,** 370–377.

Szabo, P., Shen, S., and Weksler, M. E. (1999). Age-associated defects in B lymphocyte development. *Exp. Gerontol.* **34,** 431–434.

Tarazona, R., DelaRosa, O., Alonso, C., Ostos, B., Espejo, J., Pena, J., and Solana, R. (2000). Increased expression of NK cell markers on T lymphocytes in aging and chronic activation of the immune system reflects the accumulation of effector/senescent T cells. *Mech. Ageing Dev.* **121,** 77–88.

Thompson, P. M., Hayashi, K. M., de Zubicaray, G., Scarmeas, N., Levy, G., Tang, M. X., Manly, J., and Stern, Y. (2003). Dynamics of gray matter loss in Alzheimer's disease. *J. Neurosci.* **23,** 994–1005.

Tisserand, D. J., Pruessner, J. C., Sanz Arigita, E. J., van Boxtel, M. P., Evans, A. C., Jolles, J., and Uylings, H. B. (2002). Regional frontal cortical volumes decrease differentially in aging: An MRI study to compare volumetric approaches and voxel-based morphometry. *Neuroimage* **17,** 657–669.

Trends in Health and Aging Database, U.S. Department of Health and Human Services; Centers for Disease Control and Prevention; National Center for Health Statistics.

Vaillant, G. E. (2002). "Aging Well: Surprising Guideposts to a Happier Life from the Landmark Harvard Study of Adult Development." Little, Brown, and Company, Boston.

Wechsler, D. (1958). "The Measurement and Appraisal of Adult Intelligence" 4th Ed., Williams & Wilkins, Baltimore, MD.

Weksler, M. E. (2000). Changes in the B-cell repertoire with age. *Vaccine* **18,** 1624–1628.

Weng, N. P. (2006). Aging of the immune system: How much can the adaptive immune system adapt? *Immunity* **24,** 495–499.

Whalley, L. J., Deary, I. J., Appleton, C. L., and Starr, J. M. (2004). Cognitive reserve and the neurobiology of cognitive aging. *Ageing Res. Rev.* **3,** 369–382.

Woodruff-Pak, D. S. (1997). "The Neuropsychology of Aging". Blackwell Publishers Inc., Maiden, MA.

World Health Organization (1978). Mental disorders: Glossary and guide to their classification in accordance with the ninth revision of the International Classification of Diseases. Geneva, Switzerland.

Yang, X., Stedra, J., and Cerny, J. (1996). Relative contribution of T and B cells to hypermutation and selection of the antibody repertoire in germinal centers of aged mice. *J. Exp. Med.* **183,** 959–970.

Zheng, B., Han, S., Takahashi, Y., and Kelsoe, G. (1997). Immunosenescence and germinal center reaction. *Immunol. Rev.* **160,** 63–77.

SUBCORTICAL ISCHEMIC CEREBROVASCULAR DEMENTIA

Uma Menon and Roger E. Kelley

Department of Neurology, LSU Health Sciences Center,
Shreveport, Louisiana 71103, USA

It has become increasingly apparent, especially with the advent of MRI brain scanning, that a large number of patients develop signal intensity changes in the subcortical white matter and periventricular region as they age. This appears to be accelerated by risk factors for small vessel cerebrovascular disease such as hypertension, smoking, diabetes mellitus and hyperlipidemia. The major question becomes when such changes become clinically significant. It is obvious that subcortical lacunar-type infarction can be identified by the clinical presentation. For example, typical examples of so-called "lacunar syndrome" include pure motor hemiparesis, pure sensory stroke, sensorimotor stroke, clumsy hand-dysarthria, and hemiataxia-hemiparesis. The issue becomes a measure of impact on functional ability. This is influenced by several factors. Baseline IQ and educational level, as well as expectations of age, certainly play a role. A person who develops cognitive impairment and long tract signs in their 50s or 60s is certainly going to be recognized as more impaired than an 80 year old individual who is retired and primarily is engaged in recreational activity. It would be expected that a person born with limited intellectual capacity and/or limited educational opportunity would be less likely to be identified as impaired than a person who has achieved substantial economic achievement through their innate talents.

The concept of tissue loss or lesion load becomes important when determining how pronounced the ischemic cerebrovascular changes translate into functional impairment. Correlative pathology may include cortical atrophy and ventricular

INTERNATIONAL REVIEW OF
NEUROBIOLOGY, VOL. 84
DOI: 10.1016/S0074-7742(09)00402-4

0074-7742/09 $35.00

dilatation. Loss of either cortical or subcortical tissue function is expected to be related to functional compromise. In addition, there are potential features such as the coexistence of small vessel cerebrovascular disease and Alzheimer's disease. Small vessel cerebrovascular disease might also play a contributing factor in patients susceptible to Dementia with Lewy Bodies or patients susceptible to fronto-temporal dementia or any other dementing process. Thus, the concept of tissue loss or lesion burden of disease becomes increasingly important as we recognize the potential for multifactorial issues, including genetic factors, to contribute to the phenotypic expression.

The relationships between cognitive impairment, dementia and subcortical vascular lesions are poorly understood. There have been several papers on the different aspects of cerebral insults and their impact on cognition, the various kinds of dementia and different methods of analyzing the impact of the various insults to the brain. This chapter is an attempt to review all pertinent information currently available on the poorly understood condition of "subcortical ischemic cerebrovascular dementia."

I. Terminology

It is felt that the term "Subcortical Ischemic Cerebrovascular Dementia" better describes the various categories of cognitive impairments related to ischemic changes in the brain which has been identified previously as Vascular Dementia, Vascular Cognitive Impairment or Disorder, Multi-infarct Dementia, and so on. However, there are no clear guidelines or accepted criteria for defining the complex category of SICD.

II. Various Classifications of "Vascular" Dementia

These are some of the commonly used criteria to define the dementia associated with vascular lesions in the brain (Wetterling, *et al.*, 1996):

1. DSM-IV Criteria for Vascular dementia
2. NINDS-AIREN criteria for the diagnosis of probable vascular dementia (Roman *et al.*, 1993)
3. ICD-10 research criteria (DCR-10) for dementia
4. Hachinski ischemic score

III. Cognition and Aging

Changes appear in the normal brain with aging which contribute to mental decline. It has been estimated that after the age of 65, 6.4% of patients show evidence of some cognitive impairment and this figure increases steadily with age (Ankri and Poupard, 2003). With aging, vascular lesions are seen commonly in the brain and opinions vary as to whether these lesions are clinically significant or not with regards to cognitive changes (Fig. 1). It has also been the subject of discussion about the location and volume of these lesions and whether that may have an impact on cognition.

Also with aging, the risk of vascular disease increases and the prevalence has been shown to be 0.3% during 65–69 years of age and as much as 5.2% at age 90 years (Chabriat and Bousser, 2006). The Rotterdam study showed that the prevalence of "vascular" dementia is higher in women, despite the higher incidence in men, because of increased longevity in women by 10 years (Lobo *et al.*, 2000).

Asymptomatic lacunar-type infarcts have been found to be present in 11–28% of normal elderly persons in multiple studies (Longstreth *et al.*, 1998; Price *et al.*, 1997; Vermeer *et al.*, 2003). Atrial fibrillation (Ott *et al.*, 1997) hormonal treatment (estrogen + progesterone) of women after age 65 (Shumaker *et al.*, 2003), diabetes mellitus (Leibson *et al.*, 1997) and smoking in persons above age 60 years (Juan *et al.*, 2004) have all been shown to have higher risk for "vascular" dementia.

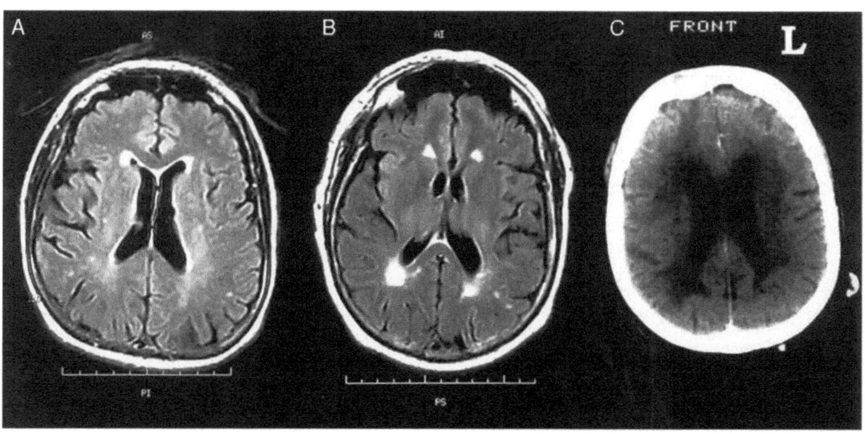

FIG. 1. Subcortical white matter hyperintensities, seen on FLAIR (transaxial) MRI brain scan, which demonstrates mild (A) and somewhat more advanced (B) periventricular findings. On noncontrast CT brain scan (C), there is prominent hypodensity in the periventricular region. These findings have been referred to as Binswanger's disease in the past, but more recently the term leukoaraiosis has been preferred because of the nonspecific clinical correlates of such findings.

Atrial fibrillation has also been shown to be an independent risk factor for cognitive impairment even in the absence of a stroke (Kilander *et al.*, 1998; O'Connell *et al.*, 1998). With the presence of hypertension, the greatest risk factor for stroke, the increased risk of cognitive impairment has been shown with different studies and patient cohorts (EVA Group, 1999; Launer *et al.*, 2000).

Stroke has been proved as a significant risk factor for cognitive impairment and dementia with as many as 25% of patients at 3 months and up to 50% within a year developing some degree of clinically significant cognitive impairment (Desmond *et al.*, 2002; Mackowiak-Cordoliani *et al.*, 2005; Tang *et al.*, 2004; Tatemichi TK *et al.*, 1994). A reported range of 25–80% of elderly subjects with dementia has Alzheimer's disease coexisting with the vascular lesions. This makes the assessment of pure cerebrovascular lesions difficult (Jellinger, 2008).

IV. Subcortical Strokes and Cognitive Impairment

It has long been believed that the small vessel or lacunar strokes in the white matter of the brain contribute minimally to the changes in cognition. However, a review of the NINDS-AIREN criteria (Roman *et al.*, 1993) demonstrates that it includes multiple basal ganglia and white matter lacunes and extensive periventricular white matter lesions on neuroimaging as causes of probable "vascular" dementia. Ischemic lesions in the brain can occur in various locations. However, with the current level and techniques of neuroimaging, it is not possible to visualize all the microinfarcts.

Areas of white matter ischemia, such as microinfarcts, often termed lacunes have been shown to be independent predictors of development of dementia (DeGroot *et al.*, 2000; Kovari *et al.*, 2004; Pasquier *et al.*, 2000; White *et al.*, 2002) particularly impairments with executive functions. Lacunes, when present in certain specific locations like the thalamus and basal ganglia, can cause cognitive disturbances (Fig. 2). Conversely, their presence in the deep white matter of the frontal, temporal, or parietal lobes does not seem to do so. Silbert *et al.* (2008) reported that an increase in total and periventricular white matter signal intensity changes, over time, correlates with progressive gait impairment while the progression of the total volume hyperintensity changes in the subcortical region is associated with memory decline in elderly subjects who are relatively intact from a cognitive standpoint.

The hippocampus is extremely sensitive to hypoperfusion which results in sclerosis. Both sclerosis and lacunes in the hippocampus can lead to decline in cognition (Bastos-Leite *et al.*, 2007; Fein *et al.*, 2000; Kril *et al.*, 2002) but this decline appears to be less than that seen in patients with Alzheimer's disease (Du *et al.*, 2002). Age appears to be the best and strongest predictor of the presence of white

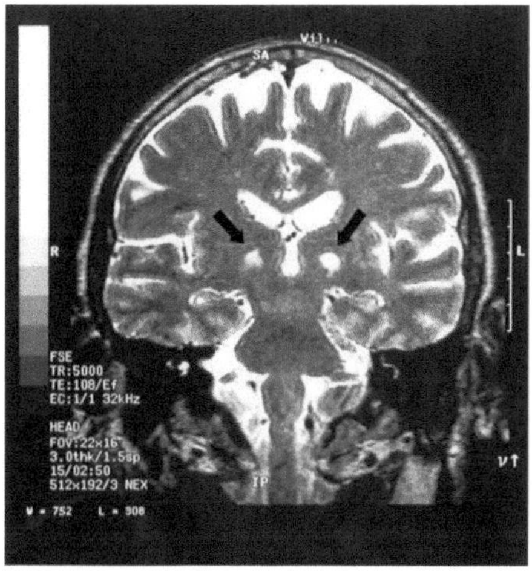

FIG. 2. Coronal T2-weighted MRI brain scan which reveals bilateral lacunar-type infarcts (arrows) in the thalamic region.

matter lesions in elderly subjects and whereas these white matter lesions maybe present in the elderly persons with normal cognitive function (Breteler *et al.*, 1994; Carey *et al.*, 2008; Liao *et al.*, 1997; Lindgren *et al.*, 1994; Longstreth *et al.*, 1996; Soderlund *et al.*, 2003; Ylikoski *et al.*, 1993), although elderly patients with dementia have been noted to have an increased load of lesions. There exists conflicting data as to whether these white matter lesions indeed contribute to cognitive decline. Some studies (Bracco *et al.*, 1993; Gold *et al.*, 2007; Schmidt *et al.*, 2002) were not able to find any definite evidence that these white matter lesions are responsible for the cognitive impairment in these subjects whereas some others including the subcortical ischemic vascular dementia (SIVD) program project (Chui, 2007; Mosley *et al.*, 2005; Prins *et al.*, 2005) have found that an increasing load of white matter lesions, over time, is associated with progressive cognitive decline (Fig. 3).

Price *et al.* (2005) report that despite the presence of white matter abnormalities, executive dysfunction is not seen until at least 25% of the white matter in the hemisphere is involved. Furthermore, when 50% of the hemisphere is involved, executive dysfunction exceeds the memory impairment. This forms the basis of research criteria proposed for subcortical vascular dementia (Erkinjuntti *et al.*, 2000). Understandably, there is no correlation between the MMSE score and white matter lesion load as the MMSE assesses the memory and cognitive functions and not the executive functions.

Carey *et al.* (2008) found that the presence of lacunes even when not producing any clinical deficits could cause decline in cognition and supports the view upheld by many others that in SIVD there is a disruption in the connecting pathways of the subcortical areas and the frontal cortex.

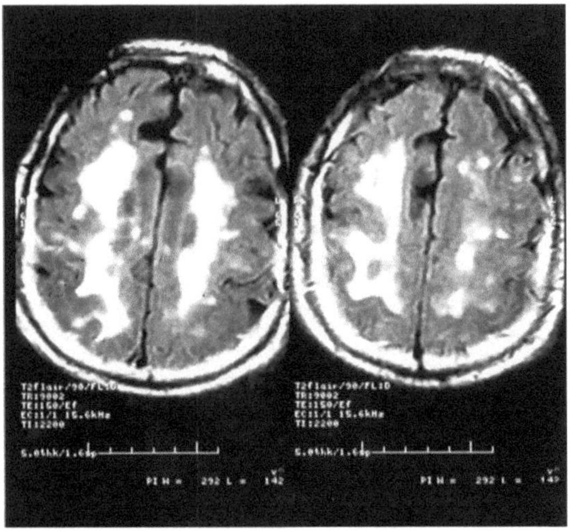

FIG. 3. Pronounced subcortical hyperintensities in the subcortical white matter, specifically the centrum semiovale, in a patient with progressive gait and cognitive impairment. This is a FLAIR (transaxial) MRI brain scan with two adjacent sections.

V. Pathophysiology of Cognitive Impairment in Subcortical Vascular Lesions

It has been proposed that various mechanisms contribute to impaired vascularity in the brain that occurs with age. Hypoperfusion related to atherosclerosis and arteriosclerosis, hypotension related to reduction in cholinergic effects, changes in the autonomic regulation of blood pressure, cortical hypometabolism, cardiac causes such as congestive heart failure with resultant systolic dysfunction and cardiac emboli and major surgical procedures have all been cited as causes for cognitive changes that occur in the older population (Baron *et al.*, 1986; Carey *et al.*, 2008; Nagata *et al.*, 2000; Roman, 2004).

The effects from the above-mentioned factors occur both in the cortical and subcortical areas. However, since the subcortical areas are mostly perfused by penetrating and deeper branches of the primary feeding arteries, the deleterious effects are most noticeable with increasing involvement of these vessels (Fig. 1). Age appears to be a significant contributing factor to the evolution of this process along with other risk factors for occlusive cerebrovascular disease. These are most readily detected as periventricular white matter lesions on neuroimaging, specifically MRI, in the elderly. It is also important to understand that all white matter lesions or hyperintensities seen are not infarcts and may potentially include causes such as astrocytic swelling as well as actual infarction (Sahlas *et al.*, 2002).

However, in a recent neuropathological study of these white matter hyperintensities (Young *et al.*, 2008), these lesions were associated with loss of vascular integrity which was interpreted to reflect a vascular origin.

It appears that any disruption in the limbic circuit, which has the anterior medial thalamus as the key component, results in memory impairment mainly involving the short term and working types (Carrera and Bogousslavsky, 2004; Swartz and Black, 2006).

Cholinergic factors have been cited as one of the possible mechanisms responsible for cognitive impairment, either causing hypotension from ineffective autonomic control related to aging or from disruption of the neurotransmitter pathways by ischemic lesions in the white matter. The treatment of Alzheimer's disease with cholinesterase inhibitors was based on the assumption that acetylcholine (Ach) is involved in memory and learning processes. Cholinergic pathways hyperintensities scale (CHIPS) developed by Bocti *et al.* (2005) provides a reliable rating scale based on white matter hyper intensities in the cholinergic pathways on MRI and their correlation to cognition in patients with Alzheimer's disease.

Whether the CHIPS scale can also be reliably used for patients with dementia from other causes as it relates to white matter ischemic lesions is not yet proved. *N*-acetyl aspartate (NAA) is a marker of neuronal integrity as measured with MRS and decreased levels are noted in the parietal and frontal cortices in subcortical ischemic cerebrovascular dementia (Capizzano *et al.*, 2000; Kattapong *et al.*, 1996; McKay *et al.*, 1996). It is thought that this correlates with the subcortical white matter lesions and supports the theory that altered metabolism plays a role in the pathogenesis.(Baron *et al.*, 1986; Nagata *et al.*, 2000) Another possible mechanism of subcortical ischemia causing cognitive deficit is interruption of the association fibers that connect the deep white matter with areas crucial for cognition (Chui, 2007). Tullberg *et al.* (2004) reported that there is an inverse relation between the glucose metabolism in the dorsolateral frontal lobe and the white matter lesion load, which supports this concept.

Altered cortical blood flow has been demonstrated with Xe-133 inhalation techniques and indicates that blood flow is primarily reduced in the ipsilateral fronto-parietal cortex with the presence of subcortical ischemic areas and reflects the subsequent changes noted in cognition and executive function (Hojer-Pederson and Friz-Peterson, 1989). Reduced volumes of the cortical gray matter and hippocampi have also been shown, by autopsy, in patients with subcortical ischemic cerebrovascular dementia (Chui, 2007; Schuff *et al.*, 2002).

Comparisons have also been drawn to the lesions of MS which also causes cognitive impairment with subcortical white matter lesions. Reduced magnetization transfer ratio (MTR) has been shown to correlate with myelin and axonal content in MS and also correlates with the level of cognitive impairment (Rovaris *et al.*, 2000, 2002; Schmierer *et al.*, 2004). Evaluation of the MTR in subcortical ischemic cerebrovascular dementia showed a reduction of MTR in the cortices (Leuchter *et al.*, 1994), again supporting the theory that remote effects in the cortices, resulting from possible disruption of the connecting tracts and circuits, may be responsible for the cognitive changes.

VI. Treatment and Prevention

Options for treatment of subcortical ischemic cerebrovascular dementia are presently quite limited. However, the overlap in pathological mechanisms between subcortical ischemic cerebrovascular dementia and Alzheimer's type dementia, with similar findings of subcortical white matter hyper intensities on MRI brain scan (Brickman *et al.*, 2008), suggests the potential for agents of some benefit in AD to be of similar benefit in subcortical ischemic dementia. Such agents include the cholinesterase inhibitors used in AD, vasodilators like nimodipine (a calcium-channel blocker) as well as some antioxidant agents. Results are somewhat favorable with the drugs used in AD and include donepezil, galantamine, and rivastigmine.

Donepezil showed significant improvement in two trials with a reduction in the cognitive decline over a period of 24 weeks (Black *et al.*, 2003; Wilkinson *et al.*, 2003) and galantamine showed some improved efficacy over placebo in another study over 12 months (Erkinjuntti *et al.*, 2002). Rivastigmine was also found to have some benefit in cognition, but this was in a relatively small open-label trial. The calcium-channel blocking agent nimodipine was tested in patients with "multi-infarct" dementia and, although initial observation appeared promising, subsequent studies were not (Pantoni *et al.*, 2000, 2005).

In view of the limited role of medications in treatment of subcortical ischemic cerebrovascular dementia, the focus is mainly on prevention. Prevention of this entity involves aggressive modification of the risk factors for stroke. Various studies have proved that control of hypertension results in a decrease in incidence of dementia up to 34% (Forette *et al.*, 1998; PROGRESS Collaborative Group, 2001; Tzourio *et al.*, 2003) and also a reduction in the incidence of stroke by as much as 35–44% (Chui, 2007). However, it is sobering to know that as many as 70% of Americans who have hypertension are unaware of their diagnosis and up to 60% are under-treated (Chobanian *et al.*, 2003). This places them at enhanced risk of stroke and other complications.

Control of other risk factors for stroke, such as diabetes mellitus, hyperlipidemia, atrial fibrillation, and so on, is also crucial for primary prevention (AHA/ASA Stroke Council, 2006). Evidence from various studies has shown the efficacy of antiplatelet agents in stroke prevention. However, there is not presently evidence-based support that the use of such agents translates into protection against dementia per se. However, by the same token, effective stroke prevention would be expected to lessen the contribution of cerebral ischemia to what might well be a multifactorial dementing process.

References

AHA/ASA Stroke (2006). Council. Primary prevention of ischemic stroke: A guideline. *Stroke* **37,** 1583–1593.

Ankri, J., and Poupard, M. (2003). Prevalence and incidence of dementia in the very old (review). *Rev. Epidemiol. Sante Publique* **51,** 349–360.

Baron, J. C., D'Antona, R., Pantano, P., Serdaru, M., Samson, Y., and Bousser, M. G. (1986). Effects of thalamic stroke on energy metabolism of the cerebral cortex. A PET study in man. *Brain* **109,** 1243–1259.

Bastos-Leite, A. J., van der Flier, W. M., van Staaten, E. C., Staekenborg, S. S., Scheltens, P., and Barkhof, F. (2007). The contribution of medial temporal atrophy and vascular pathology to cognitive impairment in vascular dementia. *Stroke* **38,** 3182–3187.

Black, S., Roman, G. C., Geldmacher, D. S., Salloway, S., Hecker, J., Burns, A., Perdomo, C., Kumar, D., and Pratt, R. (2003). Efficacy and tolerability of donezepil in vascular dementia: Positive results of a 24-week multi-center, international, randomized, placebo-controlled clinical trial. *Stroke* **34,** 2323–2330.

Bocti, C., Swartz, R. H., Gao, F-Q, Sahlas, D. J., Behl, P., and Black, S. E. (2005). A new visual rating scale to assess strategic white matter hyperintensities within cholingergic pathways in dementia. *Stroke* **36,** 2126–2131.

Bracco, L., Campani, D., Baratti, E., Lippi, A., Inzitari, D., Pracucci, G., and Amaducci, L. (1993). Relation between MRI features and dementia in cerebrovascular disease patients with leukoariosis: A longitudinal study. *J. Neurol. Sci.* **120,** 131–136.

Breteler, M. M., van Armerongen, N. M., van Swieten, J. C., Claus, J. J., Grobbee, D. E., van Gijn, J., Hofman, A., and van Harskamp, F. (1994). Cognitive correlates of ventricular enlargement and cerebral white matter lesions on MRI. The Rotterdam Study. *Stroke* **25,** 1109–1115.

Brickman, A., Honig, L. S., Scarmeas, N., Tatarina, O., Sanders, L., Albert, M. S., Brandt, J., Blacker, D., and Stern, Y. (2008). Measuring cerebral atrophy and white matter hyperintensity burden to predict the rate of cognitive decline in Alzheimer's disease. *Arch. Neurol.* **65,** 1202–1208.

Cappizzano, A. A., Schuff, N., Amend, D. L., Tanabe, J. L., Norman, D., Maudsley, A. A., Jagust, W., Chui, H. C., Fein, G., Segal, M. R., and Weiner, M. W. (2000). Subcortical ischemic vascular dementia: Assessment with quantitative MRI and H-MRS. *Am. J. Neuroradiol.* **21,** 621–630.

Carey, C. L., Kramer, J. H., Josephson, A., Mungas, D., Reed, B. R., Schuff, N., Weiner, M. W., and Chui, H. C. (2008). Subcortical lacunes are associated with executive dysfunction in cognitively normal elderly. *Stroke* **39,** 397–402.

Carrera, E., and Bogousslavsky, J. (2004). Anteromedian, central and posterolateral infarcts of the thalamus: Three variant types. *Stroke* **35,** 2826–2831.

Chabriat, H., and Bousser, M. G. (2006). Vascular dementia: Potential of antiplatelet agents in prevention. *Eur. Neurol.* **55,** 61–69.

Chobanian, A. V., Bakris, G. L., and Black, H. R., Cushman, W. C., Green, L. A., Izzo, J. L., Jr., Jones, D. W., Materson, B. J., Oparil, S., Wright, J. T., Jr., Roccella, E. J., National Heart, Lung and Blood Institute Joint National committee on Prevention, Detection, Evaluation, and Treatment of High Blood Pressure; National High Blood Pressure Eduction Program Coordinating Committee. (2003). The Seventh Report of the Joint National Committee on prevention, detection, evaluation and treatment of high blood pressure: The JNC 7 Report. *JAMA* **289,** 2560–2572.

Chui, H. C. (2007). Subcortical ischemic vascular dementia. *Neurol. Clin.* **25,** 717–740.

DeGroot, J. C., DeLeeuw, F. E., Oudkerk, M., van Gijn, J., Hofman, A., Jolles, J., and Breteler, M. M. (2002). Periventricular cerebral white matter lesions predict rate of cognitive decline. *Ann. Neurol.* **52,** 335–341.

Desmond, D. W., Moroney, J. T., Sano, M., and Stern, Y. (2002). Incidence of dementia after ischemic stroke: Results of a longitudinal study. *Stroke* **33,** 2254–2260.

Du, A. T., Schuff, N., Laakso, M. P., Zhu, X. P., Jagust, W. J., Yaffe, K., Kramer, J. H., Miller, B. L., Reed, B. R., Norman, D., Chui, H. C., and Weiner, M. W. (2002). Effects of subcortical ischemic vascular dementia and AD on entorhinal cortex and hippocampus. *Neurology* **58,** 1635–1641.

Erkinjuntti, T., Inzitari, D., Pantoni, L., Wallin, A., Scheltens, P., Rockwood, K., Roman, G. C., Chui, H., and Desmond, D. W. (2000). Research criteria for subcortical vascular dementia in clinical trials. *J. Neural. Transm. (Suppl)* **59,** 23–30.

Erkinjuntti, T., Kurz, A., Gauthier, S., Bullock, R., Lilienfeld, S., and Damaraju, C. V. (2002). Efficacy of galantamine in probable vascular dementia and AD combined with cerebrovascular disease: A randomized trial. *Lancet* **359,** 1282–1290.

EVA Group (1999). Cognitive decline in individuals with high blood pressure: A longitudinal study in the elderly. *Neurology* **53,** 1948–1952.

Fein, G., Di Sclafoni, V., Tanabe, J., Cardenas, V., Weiner, M. W., Jagust, W. J., Reed, B. R., Norman, D., Schuff, N., Greenfield, T., and Chui, H. (2000). Hippocampal and cortical atrophy predictsdementia in subcortical ischemic vascular disease. *Neurology* **55,** 1626–1635.

Forette, F., Seux, M. L., Staessen, J. A., Thijs, L., Birkenhager, W. H., Bulpitt, C. J., Girerd, X., Jaaskivi, M., Vanhanen, H., Kiviner, P., Yodfat, Y., Vanska, O., *et al.* (1998). Prevention of dementia in randomized double-blind, placebo-controlled systolic hypertension in Europe (Syst-Eur) trial. *Lancet* **352,** 1347–1351.

Gold, G., Kovari, E., Hof, P. R., Michel, J. P., Bouras, C., and Giannakopoulos, P. (2007). Sorting out the clinical consequences of ischemic lesions in brain aging: A clinicopathological approach. *J. Neurol. Sci.* **257,** 17–22.

Hojer-Pederson, E., and Friz-Peterson, O. (1989). Changes in cerebral blood flow in the cerebral cortex after subcortical ischemic infarction. *Stroke* **20,** 211–216.

Jellinger, K. A. (2008). Morphological diagnosis of "vascular dementia"—A critical update. *J. Neurolog. Sci.* **270,** 1–12.

Juan, D., Zhou, D. H., Li, J., Wang, J. Y., Gao, C., and Chen, M. (2004). A 2 year follow-up study of cigarette smoking and the risk of dementia. *Eur. J. Neurol.* **11,** 277–282.

Kattapong, V. J., Brooks, W. M., Wesley, M. H., Kodituwakku, P. W., and Rosenberg, G. A. (1996). Proton MRS of vascular and Alzheimer-type dementia. *Arch. Neurol.* **53,** 167–174.

Kilander, L., Andren, B., Nyman, H., Lind, L., Bober, M., and Lithell, H. (1998). Atrial fibrillation is an independent determinant of low cognitive function: Case-control study. *Stroke* **29,** 1816–1820.

Kovari, E., Gold, G., Herrmann, F. R., Canuto, A., Hof, P. R., Michel, J. P., Bouras, C., and Giannakopoulos, P. (2004). Cortical microinfarcts and demyelination significantly affect cognition in brain aging. *Stroke* **35,** 410–414.

Kril, J. J., Patel, S., Harding, A. J., and Halliday, G. M. (2002). Patients with vascular dementia due to microvascular pathology have significant hippocampal neuronal loss. *J. Neurol. Neurosurg. Psychiatry* **72,** 747–751.

Launer, I. J., Ross, G. W., Petrovich, H., Masaki, K., Foley, D., White, R., and Havlik, R. J. (2000). Midlife blood pressure and dementia. The Honolulu-Asia Aging Study. *Neurobiol. Aging* **21,** 49–55.

Leibson, C. L., Rocca, W. A., Hanson, V. A., Cha, R., Kokmen, E., O'Brien, P. C., and Palumbo, P. J. (1997). The risk of dementia among persons with diabetes mellitus: A population-based cohort study. *NY Acad. Sci.* **826,** 422–427.

Leuchter, A. F., Dunkin, J. J., Lufkin, R. B., Anzai, Y., Cook, I. A., and Newton, T. F. (1994). Effects of white matter disease on functional connections in the aging brain. *J. Neurol. Neurosurg. Psychiatry* **57,** 1347–1354.

Liao, D., Cooper, L., Cai, J., Toole, J., Bryan, N., Burke, G., Shahar, E., Nieto, J., Mosley, T., and Heiss, G. (1997). The prevalence and severity of white matter lesions; their relationship to age, ethnicity, gender, and cardiovascular risk factors. The ARIC Study. *Neuroepidemiology* **16,** 149–162.

Lindgren, A., Roijer, A., Rudling, O., Norrving, B., Larsson, E. M., Eskilsson, J., Wallin, L., Olsson, B., and Johansson, B. B. (1994). Cerebral lesions on MRI, heart disease and vascular risk factors on subjects without stroke. A population-based study. *Stroke* **25,** 929–934.

Lobo, A., Launer, L. J., Fritglioni, L., Anderson, K., Di Carlo, A., Breteler, M. M., Copeland, J. R., Dartigues, J. F., Jagger, C., Martinez-Lage, J., Soininen, H., and Hofman, A. (2000). for the Neurologic Diseases in the Elderly Research Group. Prevalence of dementia and major subtypes in Europe: A collaborative study of population-based cohorts. *Neurology* **54,** S4–S9.

Longstreth, W. T., Jr, Manolio, T. A., Arnold, A., Burke, G. L., Bryan, N., Jungreis, C. A., Enright, P. L., O'Leary, D., and Fried, L. (1996). Clinical correlates of white matter findings on cranial MRI of 3301 elderly people. The Cardiovascular Health Study. *Stroke* **27,** 1274–1282.

Longstreth, W. T., Jr, Bernick, C., Manolio, T. A., Bryan, N., Jungreis, C. A., and Price, T. R. (1998). Lacunar infarcts defined by MRI of 3660 elderly people. The Cardiovascular Health Study. *Arch. Neurol.* **55,** 1217–1225.

Mackowiak-Cordoliani, M. A., Bombois, S., Memin, A., Henon, H., and Pasquier, F. (2005). Post-stroke dementia in the elderly. *Drugs Aging* **22,** 483–493.

McKay, S., Meyerhoff, D. J., Constans, J-M, Norman, D., Fein, G., and Weiner, M. W. (1996). Regional gray and white matter metabolite differences in subjects with AD, with subcortical ischemic vascular dementia, and elderly controls with H MRS imaging. *Arch. Neurol.* **53,** 167–174.

Mosley, T. H., Knopman, D. S., Catellier, D. J., Bryan, N., Hutchinson, R. G., Grothues, C. A., Folsom, A. R., Cooper, L. S., Burke, G. L., Liao, D., and Szklo, M. (2005). Cerebral MRI findings and cognitive functioning. The ARIC Study. *Neurology* **64,** 2056–2062.

Nagata, K., Maruya, H., Yuya, H., Terashi, H., Mito, Y., Kato, H., Sato, M., Satoh, Y., Watahiki, Y., Hirata, Y., Yokoyama, E., and Hatazawa, J. (2000). Can PET data differentiate Alzheimer's disease from vascular dementia. *Ann. NY Acad. Sci.* **903,** 252–261.

O'Connell, J. E., Gray, C. S., French, J. M., and Robertson, I. H. (1998). Atrial fibrillation and cognitive function: Case-control study. *J. Neurol. Neurosurg. Psychiatry* **65,** 386–389.

Ott, A., Breteler, M. M., de Bruyne, M. C., van Harskamp, F., Grobbee, De, and Hofman, A. (1997). Atrial fibrillation and dementia in a population-based study. The Rotterdam Study. *Stroke* **28,** 316–321.

Pantoni, L., Bianchi, C., Beneke, M., Inzitari, D., Wallin, A., and Erkinjuntti, T. (2000). The Scandinavian Multi-Infarct Dementia Trial: A double blind, placebo-controlled trial of nimodipine in multi-infarct dementia. *J. Neurol. Sci.* **175,** 116–123.

Pantoni, L., del Ser, T., Soglian, A. G., Amigoni, S., Spadari, G., Binelli, D., and Inzitari, D. (2005). Efficacy and safety of nimodipine in subcortical vascular dementia: A randomized placebo-controlled study. *Stroke* **36,** 619–624.

Pasquier, F., Henon, H., and Leys, D. (2000). Relevance of white matter changes to pre- and post-stroke dementia. *Ann. NY Acad. Sci.* **903,** 466–469.

Price, T. R., Manolio, T. A., Kronmal, R. A., *et al.* for the Cardiovascular Health Study Collaborative Research Group. (1997). Silent brain infarction on MRI and neurological abnormalities in community-dwelling older patients. The Cardiovascular Health Study. *Arch. Neurol.* **28,** 1158–1164.

Price, C. C., Jefferson, A. L., Merino, M. D., Heilman, K. M., and Libon, D. J. (2005). Subcortical vascular dementia; integrating neuropsychological and neuroradiological data. *Neurology* **65,** 376–382.

Prins, N. D., van Djik, E. J., den Heijer, T., Vermeer, S. E., Jolles, J., Koudstaal, P. J., Hofman, A., and Breteler, M. M. (2005). Cerebral small vessel disease and decline in information processing speed, executive function and memory. *Brain* **128,** 2034–2041.

PROGRESS Collaborative Group (2001). Randomized controlled trial of perindopril-based blood pressure lowering regimen among 6105 individuals with previous stroke or TIA. *Lancet* **358,** 1033–1041.

Roman, G. C. (2004). Brain hypoperfusion: A critical factor in vascular dementia. *Neurol. Res.* **26,** 454–458.

Roman, G. C., Tatemichi, T. K., Erkinjuntti, T., Cummings, J. L., Masdeu, J. C., Garcia, J. H., Amaducci, L., Orgogozo, J. M., Brun, A., and Hofman, A. (1993). Vascular dementia: Diagnostic criteria for research studies. Report of the NINDS-AIREN International Workshop. *Neurology* **43,** 250–260.

Rovaris, M., Filippi, M., Minicucci, L., Iannucci, G., Santuccio, G., Possa, F., and Comi, G. (2000). Cortical/subcortical disease burden and cognitive impairment in patients with multiple sclerosis AJNR. *Am. J. Neuroradiol.* **21,** 402–408.

Rovaris, M., Iannucci, G., Falautano, M., Possa, F., Martinelli, V., Comi, G., and Filippi, M. (2002). Cognitive dysfunction in patients with mildly disabling relapsing-remitting multiple sclerosis: An exploratory study with diffusion tensor MR imaging. *J. Neurol. Sci.* **195,** 103–109.

Sahlas, D. J., Swartz, R. H., and Black, S. E. (2002). Clasmatodendrosis correlating with periventricular hyperintensity in mixed dementia. *Ann. Neurol.* **52,** 378–381.

Schmidt, R., Schmidt, H., Kapeller, P., Lechner, A., and Fazekas, F. (2002). Evolution of white matter lesions. *Cerebrovasc. Dis.* **13,** 16–20.

Schmierer, K., Scaravilli, F., Altmann, D. R., Barker, G. J., and Miller, D. H. (2004). Magneitization transfer ratio and myelin in postmortem MS brain. *Ann. Neurol.* **56,** 407–415.

Schuff, N., Capizzano, A. A., Du, A. T., Amend, D. L., O'Neill, J., Norman, D., Jagust, W. J., Chui, H. C., Kramer, J. H., Reed, B. R., Miller, B. L., Yaffe, K., and Weiner, M. W. (2003). Different patterns of N-acetylaspartate loss in subcortical ischemic vascular dementia and AD. *Neurology* **61,** 358–364.

Shumaker, S. A., Legault, C., Rapp, S. R., Thal, L., Wallace, R. B., Ockene, J. K., Hendrix, S. L., Jones, B. N., 3rd., Assaf, A. R., Jackson, R. D., Kotchen, J. M., Wassertheil-Smaller, J., and Wactawski-Wende, J., WHIMS Investigators. (2003). Estrogen + Progestin and the incidence of dementia and mild cognitive impairment in post-menopausal women: A randomized, controlled trial. *JAMA* **289,** 2651–2662.

Silbert, L. C., Nelson, C., Howieson, D. B., Moore, M. M., and Kaye, J. A. (2008). Impact of white matter hyperintensity volume progression on rate of cognitive and motor decline. *Neurology* **71,** 108–113.

Soderlund, H., Nyberg, L., Adolfsson, R., Nilsson, L. G., and Launer, L. J. (2003). High prevalence of white matter hyperintensities in normal aging. Relation to BP and cognition. *Cortex* **39,** 1093–1105.

Swartz, R. H., and Black, S. E. (2006). Anterior medial thalamic lesions in dementia: Frequent and volume dependently associated with sudden cognitive decline. *J. Neurol. Neurosurg. Psychiatry* **77,** 1307–1312.

Tang, W. K., Chan, S. S., Chiu, H. F., Ungvari, G. S., Wong, K. S., Kwok, T. C., Mok, V., Wong, K. T., Richards, P. S., and Ahuja, A. T. (2004). Frequency and determinants of post-stroke dementia in Chinese. *Stroke* **35,** 930–935.

Tatemichi, T. K., Paik, M., Bagiella, E., Desmond, B. W., Stern, Y., Sano, M., Hauser, A., and Mayeux, R. (1994). Risk of dementia after stroke in a hospitalized cohort: Results of a longitudinal study. *Neurology* **44,** 1885–1891.

Tullberg, M., Fletcher, E., DeCarli, C., Mungas, D., Reed, B. R., Harvey, D. J., Weiner, M. W., Chui, H. C., and Jagust, W. M. (2004). White matter lesions impair frontal lobe function regardless of their location. *Neurology* **63,** 246–253.

Tzourio, C, Anderson, C., Chapman, N., Woodward, M., Neal, B., MacMahon, S., Chalmers, J., and PROGRESS Collaborative Group. (2003). Effects of blood pressure lowering with perindopril and indapamide therapy on dementia and cognitive decline in patients with cerebrovascular disease. *Arch. Int. Med.* **163,** 1069–1075.

Vermeer, S. E., Den Heijer, T., Koudstaal, P. J., Oudkerk, M., and Breteler, M. M. (2003). Incidence and risk factors of silent brain infarcts in population-based Rotterdam scan study. *Stroke* **34,** 392–396.

Wetterling, T., Kanitz, R. D., and Borgis, K. J. (1996). Comparison of different diagnostic criteria for vascular dementia (ADDTC, DSM-IV, ICD-10, NINDS-AIREN). *Stroke* **27,** 30–36.

White, L., Petrovich, H., Hardman, J., Launer, L. J., Curb, J. D., and Schmidt, R. (2002). Cerebrovascular pathology and dementia in autopsied Honolulu Aging Study participants. *Ann. NY Acad. Sci.* **997,** 9–23.

Wilkinson, D., Doody, R., Taubman, K., Mintzer, J., Kertesz, A., Pratt, R. D., and Donepezil 308 Study Group. (2003). Donepezil in vascular dementia: A randomized, placebo-controlled study. *Neurology* **61,** 479–486.

Ylikoski, R., Ylikoski, A., Erkinjuntti, T., Suldava, R., Raininko, R., and Tilvis, R. (1993). White matter changes in healthy elderly persons correlate with attention and speed of mental processing. *Arch. Neurol.* **50,** 818–824.

Young, V. G., Halliday, G. M., and Kril, J. J. (2008). Neuropathologic correlates of white matter hyperintensities. *Neurology* **71,** 804–808.

CEREBROVASCULAR AND CARDIOVASCULAR PATHOLOGY IN ALZHEIMER'S DISEASE

Jack C. de la Torre

Sun Health Research Institute, Center for Alzheimer's Research, Sun City, Arizona 85351, USA

Presently, compelling evidence indicates that Alzheimer's disease (AD) is a vascular disorder with neurodegenerative consequences and should be treated as such. A substantial body of evidence including epidemiological, pharmacotherapeutic, and neuroimaging studies support the concept of AD as a vascular disorder or vasocognopathy that initiates as brain hypoperfusion, creating a neuronal energy crisis. The neuronal energy crisis provokes the cellular and molecular changes that characterize this dementia. In this brief review, the many vascular-related risk factors to AD including some that are preventable or amenable to treatment are discussed. Considerable human data now point in a new direction for guiding future research into AD. This new research direction should open a window of opportunity for decisive management and treatment of this devastating disorder.

I. Introduction

Cerebral energy supply can be interrupted after cardiac arrest or stroke, two conditions that result in global or local brain cell death. However, brain energy supply can dwindle slowly over long periods of time following non-fatal carotid artery stenosis or cardiac disease, two major disorders that result in brain hypoperfusion. Brain hypoperfusion (not an ischemic process) does not necessarily kill or damage nerve cells right away by significantly arresting energy nutrients delivered via the circulation but may chronically "gnaw" at nerve cells over a long period of time by destabilizing their normal function (Fig. 1).

INTERNATIONAL REVIEW OF NEUROBIOLOGY, VOL. 84
DOI: 10.1016/S0074-7742(09)00403-6

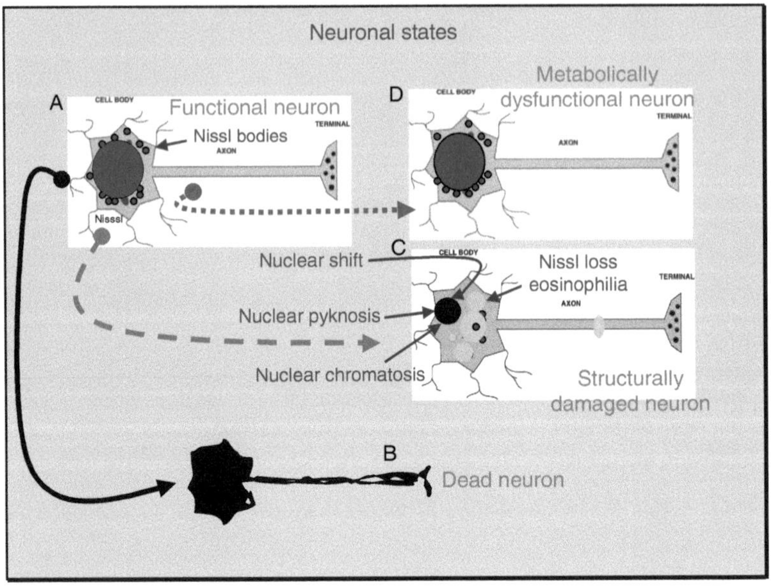

FIG. 1. Functional and pathological states deranging brain neurons resulting in degenerative processes. Healthy, functional neurons (A) can either (B) die (solid arrow), (C) undergo metabolic-structural damage (dashed arrow), (D) undergo metabolic only damage (dotted arrow). In the (C) state, intracellular damage can be assessed histologically by noting nuclear or cytoplasmic changes in organelles (chromatolysis, eosinophilia, nuclear shift, pyknosis, etc.). In the (D) state, bio-molecular markers reflecting neuronal distrenss are needed to evaluate extent of metabolic dysfunction. Neuronal metabolic dysfunction following chronic brain hypoperfusion is likely the first step in the pathologic pathway to cytostructural damage, atrophy, and death and how such dysfunction may explain cognitive disability, as in pre-clinical Alzheimer's disease.

Since cerebral hypoperfusion now appears to be a major precursor of Alzheimer's disease (AD) (Aguero-Torres *et al.*, 2006; Breteler, 2000; de la Torre, 2000, 2004a,b; de la Torre and Mussivand, 1993), the issue of glucose and oxygen availability and delivery via the circulation becomes a key element in the puzzle to understand the cellular-molecular cascade that leads to AD. The main reason to focus on glucose uptake and utilization by brain cells is that this molecule is the primary energy substrate for the brain which also consumes 20% of the total oxygen supply used by the body (Fig. 1A).

We became interested in the role of cerebral energy metabolism on AD in the early 1990s, when the subject was only of cursory, rare interest, following our series of experimental brain hypoperfusion studies and a review of the clinical dementia literature at the time (de la Torre and Mussivand, 1993). These experimental studies led us to propose the vascular hypothesis of AD as the likely cause of this disorder (de la Torre and Mussivand, 1993).

Since then, compelling evidence from a variety of clinical and basic studies has convince us that AD should be treated as a vasocognopathy (de la Torre, 2004) (a vascular-related, cognitive disorder) and we shall explore some of this evidence in our review here to determine whether proof of concept is met.

II. Risk Factors to AD

The most compelling evidence for a vascular connection to AD comes from independent epidemiological studies. Findings from the Rotterdam Study, the Kungsholmen project, EURODEM, FINMONICA, the PROCAM study, the Framingham study and the Honolulu-Asia study, among others, indicate that over two dozen risk factors, all vascular related, have thus far been recognized for AD (Table I) (Breteler, 2000; Aguero-Torres *et al.*, 2006; de la Torre and Mussivand, 1993). These epidemiological studies have led to the conclusion that most of the AD cases analyzed have a vascular involvement, and that pure dementia types (including vascular dementia) in older subjects constitute only a minority of dementia cases (Aguero-Torres *et al.*, 2006; Breteler, 2000). It is highly improbable that chance alone could explain the reduction of cerebral blood flow (CBF) induced by several dozen different risk factors found linked to AD (Table I). If indeed, a variety of vascular risk factors for AD exists and is highly important in promoting this dementia, then the question of the role of amyloid beta ($A\beta$) needs to be asked. Do risk factors eventually also promote the excessive production of $A\beta$ in brain or does $A\beta$ promote the varied risk factors? The second possibility seems rather unlikely due to the spectral nosology of each risk factor described thus far (Table I), so building on the assumption that vascular-dependent risk factors for AD *do* lower cerebral perfusion and eventually promote $A\beta$ production, we have introduced the role of *critically attained threshold of cerebral hypoperfusion* (CATCH) (de la Torre, 2000). CATCH develops when a vascular risk factor or factors further diminishes cerebral CBF in an individual who already has developed declining cerebral perfusion due to advanced age and a poor vascular reserve. The reason is that CBF normally declines with age by about 21% from age 21 to 60 (Leenders *et al.*, 1990). Below this normal-age related CBF decline, a "CATCH brain blood flow level" can result from advancing age when it co-exists with one or more vascular risk factor as listed in Table I.

The speed at which CATCH is attained depends on the person's state of health, lifestyle, genetics, gender, diet, environmental exposure, and other confounding factors. Once CATCH is activated, energy hypometabolism and oxidative stress result from continuous brain hypoperfusion and from reduced glucose and oxygen delivery to neurons and glia (de la Torre, 2000) (Fig. 1B). The main reason for the high glucose/O_2 uptake in the normal brain is the massive amounts of ATP needed

TABLE I

SUMMARY OF FINDINGS FROM EPIDEMIOLOGICAL STUDIES (SEE TEXT) OF VASCULAR-RELATED RISK FACTORS TO ALZHEIMER'S DISEASE

Alzheimer's disease vascular risk factors
Brain-related risk factors
Aging
Ischemic stroke
Silent stroke
Head injury
Transient ischemic attacks
Migraine
Lower education
Hemorheologic abnormalities
Depression
Circle of Willis atherosclerosis
Heart-related risk factors
Congestive heart failure
Valvular disease
Hypertension
Hypotension
Thrombotic episodes
High serum homocysteine
Atrial fibrillation
Presence of ApoEε4 allele
Carotid atherosclerosis
Coronary artery bypass surgery
Peripheral risk factors
High serum cholesterol
High intake of saturated fat
Diabetes mellitus II
Hemorheologic abnormalities
Alcoholism
Smoking
Menopause

Highly prevalent conditions such as stroke, hypertension, atherosclerosis and heart disease are common precursors to AD.

to maintain neuronal signaling during active ion channel flux. Ion flux is the basis for propagation of action potentials and neurotransmission. Impaired brain energy production also upregulates the Aβ rate-limiting enzyme BACE1, thus promoting Aβ overproduction (Velliquette *et al.*, 2005) (Fig. 1B). Consequently, when mitochondrial function in neurons is disrupted by failing glucose/O_2 supply for energy production, rapid damage to brain cells occurs. Brain hypoperfusion is the pathogenic trigger that pushes the neuronal energy crisis and the cascade of molecular and cytopathologic changes that characterize AD (Fig. 2).

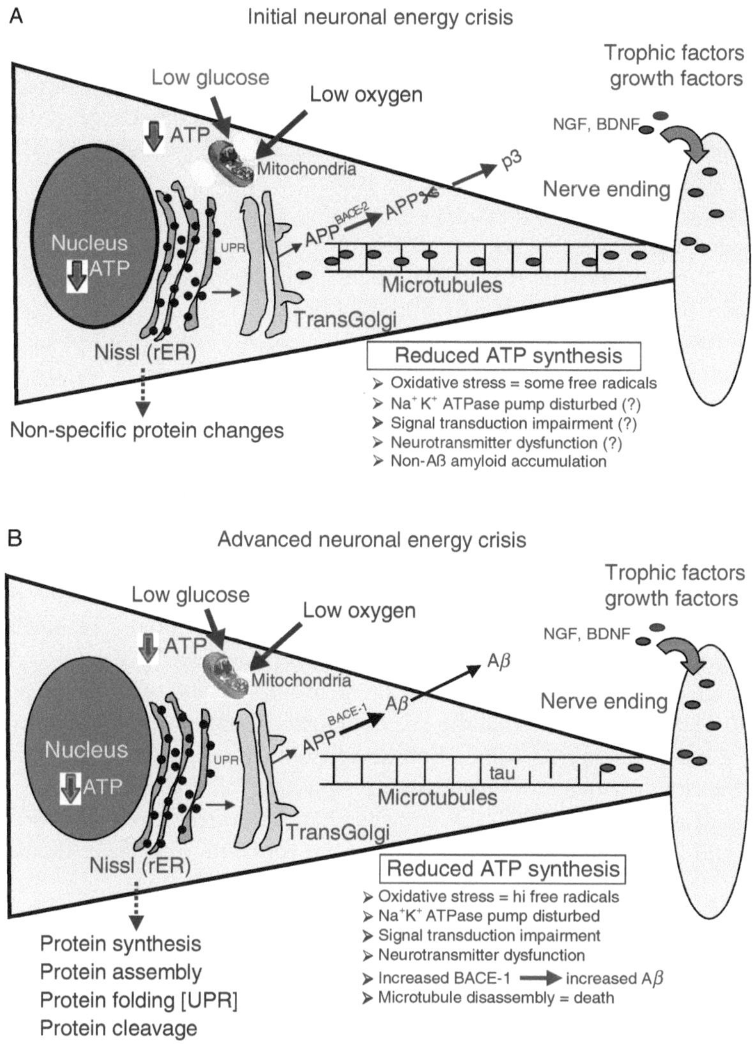

FIG. 2. (A) A fundamental principle in cell biology is seen by the use of chemical energy in the form of ATP derived from glucose/oxygen delivery to assemble, disassemble and alter protein structure. Since proteins work to keep a neuron healthy, defects in their synthesis, assembly, folding, or cleavage can impair the normal intracellular and extracellular secretory transport pathway and damage brain cells. Normal neuronal energy metabolism in the brain includes optimal ATP production by mitochondrial oxidative phosphorylation, RNA transcription and protein synthesis, cell signaling, neurotransmission, axonal transport of molecules between cytoplasm and nerve ending via microtubules. Normal brain has little or no Aβ accumulation. Following initial cerebral hypoperfusion, reduced ATP synthesis can result in non-specific protein and oxidative stress changes that may lead to MCI. Theoretically, overexpression of Aβ peptides and specific neurodegenerative pathology may be lacking.

Risk factors for AD are either the result of genetic susceptibility (e.g., carrying an ApoE ε4 allele) or environmental exposure of a person's health to an event that can introduce, accelerate, or further compromise cognitive dysfunction. At the present time, only environmental risk factor exposure is modifiable or preventable but in the future, genetic engineering of potential risk susceptibilities to AD could become a realistic therapeutic target for AD.

Vascular risk factors to AD as shown in Table I have been extensively reviewed (Aguero-Torres *et al.*, 2006; Breteler, 2000; de la Torre and Mussivand, 1993; Sjogren *et al.*, 2006; Sparks *et al.*, 2005) and only some recently reported risk factors will be briefly discussed here.

One example of a potentially modifiable vascular risk to AD of increasing research interest is hypercholesterolemia and dietary fat intake. A steady amount of research has confirmed a link between high levels of cholesterol and the development of AD (Sjogren *et al.*, 2006).

Hypercholesterolemia acts as a precursor of atherosclerosis, cardiovascular disease, and diabetes. For obvious reasons, high serum cholesterol levels have generated a rapid-growing market for lipid-lowering drugs prescribed to patients at risk of cardiovascular or cerebrovascular conditions. Statins that lower cholesterol levels have been suggested as useful in both prevention and treatment of AD but this conclusion needs further verification (Sparks *et al.*, 2005). The mechanism by which statins may provide a benefit against cerebrovascular disease including AD remains speculative and is likely multifactorial. In 1998, it was suggested, on the basis of murine studies, that the health benefits of statins did not merely aim at lowering lipids (Laufs *et al.*, 1998). Statins were shown to protect against injury by a mechanism involving the selective upregulation of endothelial nitric oxide synthase (Laufs *et al.*, 1998), the enzyme that generates nitric oxide in blood vessels and is involved with regulating vascular function. Further research along

(B) Persistent reduction of glucose and oxygen delivery to ischemic-sensitive neurons following *chronic* brain hypoperfusion leads to critical ATP depletion and promotes oxidative stress, reactive oxygen species and abnormalities in protein synthesis activating their disassembly, misfolding (generating unfolded protein response, UPR), and abnormal cleavage. Altered protein synthesis can subsequently form molecules that can ravage nucleic acids, protein kinases, phosphatases, lipids, and harm cell structure. ATP cutback also hastens ionic pump dysfunction, signal transduction breakdown, neurotransmitter failure, faulty cleavage of amyloid precursor protein (APP) leading to BACE-1 upregulation and Aβ overproduction and microtubule damage from tau hyperphosphorylation. Retrograde axonal transport (curved arrow) of trophic and growth factors (e.g., NGF, BDNF) essential for neuronal survival is diminished due to energy-starved motor protein deficiency. This *selective neuronal* energy crisis is caused by *blood flow supply not meeting cell energy demand* in highly active brain regions whose vascular reserve capability has reached a critical threshold. About 75% of ATP energy is used on signaling (action potentials, postsynaptic potentials, etc.). Reduced ATP synthesis is the precursor of the molecular cascade that leads to Alzheimer's disease (AD) and to the region-specific neurono-glial death pathway. Key: ↓ = reduction or loss; ⇣ = impaired reaction.

these lines has demonstrated that statins improve endothelial function, increase nitric oxide bioavailability, provide antioxidant properties, inhibit inflammatory responses, exert immunomodulatory actions, regulate progenitor cells, and stabilize atherosclerotic plaques (Endres *et al.*, 1998) and several studies have shown a benefit for statins in the treatment of AD (Wolozin *et al.*, 2000).

Another important modifiable risk factor to AD is metabolic syndrome (MeS). MeS is a cluster of factors that can lead to the development of cardiovascular disease. MeS is characterized by abnormal insulin, glucose, and lipoprotein metabolism as well as by hypertension and obesity (Grundy *et al.*, 2005). These factors are frequently found in people with either excess abdominal body fat or a decreased ability of the body to use insulin, which is known as insulin resistance (Grundy *et al.*, 2005).

MeS has been found to be associated with AD even when diabetic patients are excluded from analysis (Vanhanen *et al.*, 2006), suggesting that obesity associated with cardiovascular dysfunction or hypercholesterolemia is sufficient to initiate cognitive impairment that later can convert to AD. The number of people with MeS increases with age and is estimated to affect 40% of people beyond the age of 60. People with uncontrolled diabetes mellitus type 2 and those with heart disease or prone to stroke are most likely to develop MeS (Martins *et al.*, 2006).

Since obesity and lack of exercise are two conditions that can lead to MeS, correcting these by lifestyle changes that include a healthy diet and physical activity could help prevent or reverse MeS. The cardiovascular risk factors associated with MeS contribute to the development of atherosclerosis that can lead to a heart attack or a stroke. People with the MeS are also more likely to develop type 2 diabetes mellitus (Martins *et al.*, 2006). Insulin resistance is central to type 2 diabetes and is also implicated in the pathogenesis of AD (Vanhanen *et al.*, 2006). This has prompted ongoing clinical trials in AD patients to test the efficacy of improving insulin-like signaling with dietary omega-3 fatty acids or insulin-sensitizing drugs as well as by exercise regimens.

It is well established that hypertension can increase the risk of stroke and heart problems and decrease life expectancy (de la Torre, 2006). Many studies including the Framingham and the Honolulu-Asia Aging studies have implicated impaired cognitive function to hypertension in geriatric patients (Elias *et al.*, 1993). It has also been known for some time that hypertension in the elderly is a potential risk factor to AD (de la Torre, 2009). What is not clear is how hypertension increases the incidence of AD, particularly in those not-treated with antihypertensives. People with hypertension are six times more likely to have a stroke that can lead to AD (Ivan *et al.*, 2004). It is calculated that reducing the number of cases affected by stroke which are amenable to prevention or treatment, would have a major impact in lowering the incidence and societal costs of Alzheimer's disease which is now estimated to affect 25 million people worldwide.

A. Cardiovascular Disease

Some forms of cardiac disease that result in impairment of cardiac output or function with consequent reduction of cerebral perfusion have been found to increase AD risk (Alves and Busatto, 2006).

Since 20% of cardiac output goes to the brain and 80% of carotid artery flow goes to each ipsilateral middle cerebral artery, it is no surprise that vascular lesions of the gray and white matter are frequently associated with cardiac disease and carotid artery stenosis, both vasculopathies that can induce cognitive impairment and probably AD (Alves and Busatto, 2006; de Leeuw *et al.*, 2004; Ivan *et al.*, 2004; Miklossy, 2003).

Moreover, cerebral hypoperfusion can result in white matter lesions and cortical watershed microinfarcts in the absence of atherosclerosis or amyloid angiopathy (Miklossy, 2003), and may induce microinfarcts in the hippocampus prior to the onset of AD (de Leeuw *et al.*, 2004).

We firmly believe that AD onset can be delayed or possibly prevented if the heart can be kept healthy and brain hypoperfusion can be treated by restoring normal blood flow. The upshot would be a major research breakthrough in AD prognosis if either feat is successfully achieved. Consequently, diet, exercise, and management of many vascular risk factors amenable to treatment may be the key to significantly reducing the incidence of this dementia in the future.

B. Preclinical Detection of AD

The most important factor in developing a useful blueprint for AD treatment is identifying such candidates at the *preclinical* stage before any incapacitating cognitive damage has occurred. The reason is that once cognitive meltdown has begun (which implies serious neuronal damage and tissue atrophy), AD pathology may be most difficult to reverse or arrest.

Considering AD as a vascular disorder characterized by brain hypoperfusion, techniques that can detect regional CBF reduction during mild cognitive impairment (MCI), a condition often seen prior to AD, or even before any memory deficits are measured, will be crucial to predict AD candidates.

In our judgment, three of the most important neuroimaging tools to screen AD candidates before major clinical symptoms have developed are single-photon emission tomography (SPECT), injected [18F]fluorodeoxyglucose (FDG) using positron emission tomography (PET), and transcranial Doppler (TCD) ultrasonography. SPECT and TCD can detect with a good degree of sensitivity and specificy, persons showing regional brain hypoperfusion at the MCI stage, many who later convert to AD (Johnson *et al.*, 1998; Ruitenberg *et al.*, 2005).

Preclinical detection of AD as a group is now possible using FDG-PET although the technique is considerably more expensive and labor intense than SPECT or TCD. Individuals at risk of AD but who do not express any cognitive deficits show local hypometabolic reductions of the cerebral metabolic rate of glucose after FDG in brain regions that will later develop atrophy and neurodegeneration (de Leon *et al.*, 2001). Since measures of glucose metabolic rate and CBF are coupled (Silverman and Phelps, 2001), FDG-PET studies are proof of concept that regional brain hypoperfusion precedes Alzheimer clinical symptoms, Aβ production, and neurodegenerative changes. FDG-PET can also track AD progression (de Leon *et al.*, 2001).

Existing or developing coronary artery disease in the elderly patient may worsen age-related cerebral hypoperfusion in the elderly person (de la Torre, 2006). Most patients with ischemic heart disease will present themselves in general practice. Therefore, the community management of ischemic heart disease has already become increasingly important and the role of the primary care physician even more pivotal.

Inasmuch as normalization of diminished cardiac output from any cause may prevent or reverse MCI (a high risk for prodromal AD) (de la Torre, 2006, 2008, 2009), simple, safe, and reliable diagnostic screens should be applied to older patients in an effort to reduce the consequences of cardiac-to-brain related AD risk factors. We have recommended the use of cost-effective, safe, and reliable noninvasive tools such as echocardiography (Fig. 3) and carotid Doppler ultrasound to be used in the elderly individual found to complain of memory difficulties (de la Torre, 2008). Employment of noninvasive screening techniques using carotid artery ultrasound and echocardiography (CAUSE) (de la Torre, 2008, 2009) in the detection of medically or surgically correctable conditions that reduce heart or carotid artery blood flow to brain is anticipated to delay or prevent the progression to Alzheimer's or vascular dementia (VaD) in many of those affected. These office procedures involve no needles, no radiation, and no contrast agents that may cause allergic reactions.

Carotid artery ultrasound is a safe, cost-effective, and reasonably reliable way to determine the clinical action that may prevent potential or continuing cognitive decline. Such clinical action could significantly impact on the prevalence of incident dementia rate. Echocardiographic measurements can provide a relatively simple, safe, reliable, and cost-effective evaluation of impaired cardiac function that may indicate a dangerously reduced level of cerebral perfusion that can lead to AD.

Echocardiography is also useful in guiding therapy over time. This is possible by measuring the effect of therapy on multiple parameters of cardiac function, including diastolic filling and ventricular performance, two major determinants of preload and therefore of cardiac output (Ahmed *et al.*, 2007). These measurements have demonstrated considerable prognostic value in symptomatic and

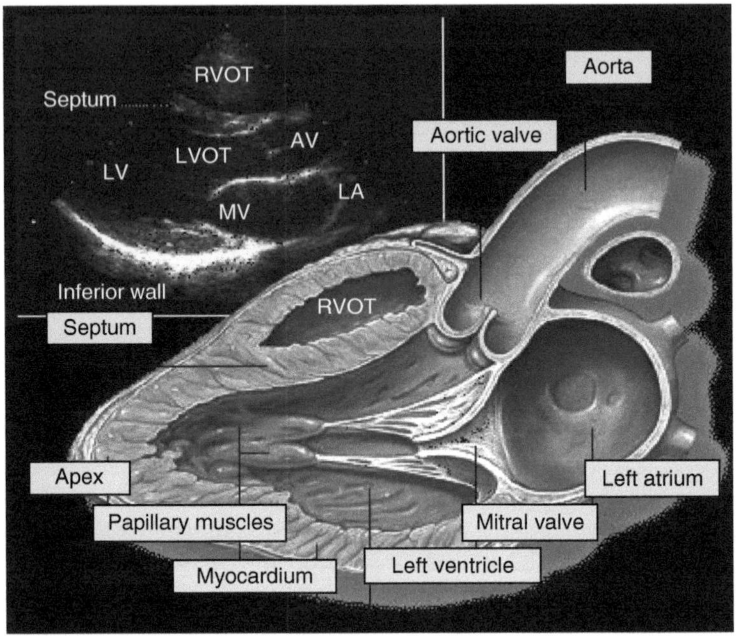

FIG. 3. Echocardiography image (top) is obtained by placing the ultrasound probe on the surface of the chest over the heart. Image shows all four chambers of the heart, useful for detecting chamber enlargement. Valvular disease, hypertension, arrhythmias, left ventricular function, ejection fraction, and cardiac output can be determined and measured. These cardiac measurements offer a noninvasive, cost-effective and reliable approach that can identify often correctable or treatable early lesions of the heart which may contribute to chronic brain hypoperfusion and possibly AD in the elderly patient with mild memory complaints. Key: RVOT = right ventricular outlet tract; LVOT = left ventricular outlet tract; LV = left ventricle; MV = mitral valve; LA = left atrium. (See Color Insert.)

asymptomatic patients with either preserved or abnormal cardiac function (Ahmed *et al.*, 2007). Such techniques can generally rule out cardiac pathology and extracranial vessel disease. Cardiac pathology and extracranial vessel disease have been shown to be major contributors to cerebral hypoperfusion and stroke (Alves and Busatto, 2006) and their attrition may result in a significant impact on lowering the prevalence of Alzheimer's and VaD.

C. Treatment of AD

There is presently no effective treatment for AD. Five drugs are available in the US market for prescriptive use in AD: Cognex (tacrine), Aricept (donepezil), Excelon (rivastigmine tartrate), Reminyl (galantamine hydrobromide), and

Namenda (memantine). The first four above act to slow the synaptic breakdown of acetylcholine, one of the neurotransmitters important in memory and learning. Reminyl also targets both AD and "mixed" dementia, that is, AD complicated by cerebrovascular pathology.

A fifth drug, memantine, is a relatively newer medication for the treatment of Alzheimer's disease that works ostensibly by antagonizing the *N*-methyl-D-aspartate receptors. Nonprescriptive treatments for AD in present use are nonsteroidal anti-inflammatory agents (NSAIDs), ginkgo biloba, estrogen, vitamin E, and prescriptive statins. All these treatments, whether prescriptive or over-the-counter, at best, provide only very modest control of symptoms generally at the early stages of AD and offer little to no cost-benefit ratio at the later stages of the disease (Loveman *et al.*, 2006). Their modest benefit to AD patients must be weighed against their frequent side-effects, particularly the anti-cholinergic agents which include nausea, vomiting, dizziness, headaches, and depression. Moreover, since most clinical trials on the effectiveness of these medicines were sponsored by industry or industry-paid researchers, a prominent shadow for positive bias has to be considered (Sismondo, 2008).

It should be noted that the common link binding practically all the medicines and therapies available or tested experimentally for AD so far is their ability to *mildly* increase cerebral perfusion (de la Torre, 2004b). However, since cerebral perfusion is only slightly improved, the beneficial effect of medicines such as those listed above is transient and limited. Interestingly, most of the products that improve AD symptoms also improve the symptoms of VaD. This activity by medicaments with differing pharmacokinetics and pharmacodynamics, lend additional support to the concept of brain hypoperfusion in AD and VaD and reinforce potential therapeutic targets aimed at patients who initially present with mild cognitive dysfunction and who display declining perfusion to the brain. New technologies, whose primary goal is to significantly increase cerebral perfusion in such patients, could delay, reverse, or prevent the neuropathological process that can lead to dementia.

III. Conclusions

There is now compelling evidence that Alzheimer's disease is a vascular disorder with neurodegenerative consequences and should be treated as such. A large body of evidence including epidemiological, pharmacotherapeutic, and neuroimaging studies support the concept of AD as a vascular disorder (a vasocognopathy; de la Torre, 2004a) that results in brain hypoperfusion and consequent neuronal energy crisis that provokes the cellular and molecular changes that characterize this dementia. Cerebrovascular and cardiovascular conditions responsible for reducing carotid artery and cardiac blood flow to the brain may be crucial clues in better understanding the physiopathology of AD. This shift in thinking acknowledges the notion that current complacency on the state of the art has created the gridlock which ignores that something is wrong in our ability to clinically manage AD after a century of research. In this brief review, the many vascular-related risk factors to AD including some that are preventable or amenable to treatment are discussed. Considerable human data now point in a new direction for guiding future research into AD. A more data driven new research direction should open a window of opportunity that can lead to new technologies, create new subfields by revolutionizing the field of dementia and provide decisive management of this devastating disorder.

Acknowledgment

This research was supported by an Investigator-Initiated Research Grant from the Alzheimer's Association.

References

Aguero-Torres, H., Kivipelto, M., and von Strauss, E. (2006). Rethinking the dementia diagnoses in a population-based study: What is Alzheimer's disease and what is vascular dementia? A study from the Kungsholmen project. *Dement. Geriatr. Cogn. Disord.* **22**(3), 244–249.

Ahmed, S. N., Syed, F. M., and Porembka, D. T. (2007). Echocardiographic evaluation of hemodynamic parameters. *Crit. Care Med.* **35**(8 Suppl.), S323–S329.

Alves, T. C., and Busatto, G. F. (2006). Regional cerebral blood flow reductions, heart failure and Alzheimer's disease. *Neurol. Res.* **28,** 579–587.

Alves, T. C., and Busatto, G. F. (2006). Regional cerebral blood flow reductions, heart failure and Alzheimer's disease. *Neurol. Res.* **28**(6), 579–587.

Breteler, M. M. (2000). Vascular risk factors for Alzheimer's disease: An epidemiologic perspective. *Neurobiol. Aging* **21**(2), 153–160.

de la Torre, J. C., and Mussivand, T. C. (1993). Can disturbed brain microcirculation cause Alzheimer's disease? *Neurol. Res.* **15,** 146–153.

de la Torre, J. C. (2000). Critically attained threshold of cerebral hypoperfusion: The CATCH hypothesis of Alzheimer's pathogenesis. *Neurobiol. Aging* **903,** 424–436.

de la Torre, J. C. (2004a). Alzheimer's disease is a vasocognopathy: A new term to describe its nature. *Neurological. Res.* **26**(5), 517–524.

de la Torre, J. C. (2004b). Is Alzheimer's a neurodegenerative or a vascular disorder? Data, dogma and dialectics. *Lancet Neurol.* **3,** 184–190.

de la Torre, J. C. (2006). How do heart disease and stroke become risk factors for Alzheimer's disease? *Neurol. Res.* **28,** 637–644.

de Leeuw, F. E., Barkhof, F., and Scheltens, P. (2004). White matter lesions and hippocampal atrophy in Alzheimer's disease. *Neurology* **62,** 310–312.

de la Torre, J. C. (2008). Alzheimer disease prevalence be lowered with non-invasive testing. *J. Alzheimer Dis.* **14**(3), 353–359.

de la Torre, J. C. (2009). Carotid artery ultrasound and echocardiography (CAUSE) testing to lower the prevalence of Alzheimer's disease. *J. Stroke Cerebrovasc. Dis.,* in press.

de Leon, M. J., Convit, A., and Wolf, O. T. (2001). Prediction of cognitive decline in normal elderly subjects with 2-[18F]fluoro-2-deoxy-d-glucose/positron-emission tomography (FDG/PET). *Proc. Natl. Acad. Sci. USA* **98,** 10966–10971.

Elias, M. F., Wolf, P. A., D'Agostino, R. B., Cobb, J., and White, L. R. (1993). Blood pressure level is inversely related to cognitive functioning: The Framingham Study. *Am. J. Epidemiol.* **138**(6), 353–364.

Endres, M., Laufs, U., and Huang, Z. (1998). Stroke protection by 3-hydroxy-3-methylglutaryl (HMG)-CoA reductase inhibitors mediated by endothelial nitric oxide synthase. *PNAS* **95,** 8880–8885.

Grundy, S. M., Cleeman, J. I., Daniels, S. R., and Donato, K. A. (2005). Diagnosis and management of the metabolic syndrome: An American Heart Association/National Heart, Lung, and Blood Institute Scientific Statement. *Circulation* **112,** 2735–2752.

Ivan, C. S., Seshadri, S., and Beiser, A. (2004). Dementia after stroke: The Framingham Study. *Stroke* **35,** 1264–1268.

Johnson, K. A., Jones, K., Holman, B. L., Becker, J., Spiers, P. A., Satlin, A., and Albert, M. S. (1998). Prediction of Alzheimer's disease using SPECT. *Neurology* **50,** 1563–1571.

Laufs, U., La Fata, V., Plutzky, J., and Liao, J. K. (1998). Upregulation of endothelial nitric oxide synthase by HMG CoA reductase inhibitors. *Circulation* **97**(12), 1129–1135.

Leenders, K. L., Perani, D., Lammertsma, A. A., Heather, J. D., Buckingham, P., Healy, M. J., Gibbs, J. M., Wise, R. J., Hatazawa, J., and Herold, S. (1990). Cerebral blood flow, blood volume and oxygen utilization. Normal values and effect of age. *Brain* **113,** 27–47.

Loveman, E., Green, C., Kirby, J., Takeda, A., Picot, J., Payne, E., and Clegg, A. (2006). The clinical and cost-effectiveness of donepezil, rivastigmine, galantamine and memantine for Alzheimer's disease. *Health Technol. Assess* **10**(1):iii-iv, ix-xi, 1–160.

Martins, I. J., Hone, E., Foster, J. K., Sunram-Lea, S. I., Gnjec, A., Fuller, S. J., Nolan, D., Gandy, S. E., and Martins, R. N. (2006). Apolipoprotein E, cholesterol metabolism, diabetes, and the convergence of risk factors for Alzheimer's disease and cardiovascular disease. *Mol. Psychiatry* **11**(8), 721–736.

Miklossy, J. (2003). Cerebral hypoperfusion induces cortical watershed microinfarcts which may further aggravate cognitive decline in Alzheimer's disease. *Neurol. Res.* **25**(6), 605–610.

Ruitenberg, A., den Heiker, T., and Bakker, S. L. (2005). Cerebral hypoperfusion and clinical onset of dementia: The Rotterdam study. *Ann. Neurol.* **57,** 789–794.

Sjogren, M., Mielke, M., Gustafson, D., Zandi, P., and Skoog, I. (2006). Cholesterol and Alzheimer's disease—is there a relation?. *Mech. Ageing Dev.* **127**(2), 138–147.

Silverman, D. H., and Phelps, M. E. (2001). Application of positron emission tomography for evaluation of metabolism and blood flow in human brain: Normal development, aging, dementia, and stroke. *Mol. Genet. Metab.* **74**(1-2), 128–138.

Sismondo, S. (2008). How pharmaceutical industry funding affects trial outcomes: Causal structures and responses. *Soc. Sci. Med.* **66**(9), 1909–1914.

Sparks, D. L., Sabbagh, M. N., and Connor, D. J. (2005). Atorvastatin for the treatment of mild to moderate Alzheimer disease: Preliminary results. *Arch. Neurol.* **62,** 753–757.

Vanhanen, M., Koivisto, K., Moilanen, L., Helkala, E. L., Hanninen, T., Soininen, H., Kervinen, K., Kesaniemi, Y. A., Laakso, M., and Kuusisto, J. (2006). Association of metabolic syndrome with Alzheimer disease: A population-based study. *Neurology* **67**(5), 843–847.

Velliquette, R. A., O'Connor, T., and Vassar, R. (2005). Energy inhibition elevates beta-secretase levels and activity and is potentially amyloidogenic in APP transgenic mice: Possible early events in Alzheimer's disease pathogenesis. *J. Neurosci.* **25,** 10874–10883.

Wolozin, B., Kellman, W., Ruosseau, P., Celesia, G., and Siegel, G. (2000). Decreased prevalence of Alzheimer disease associated with 3-hydroxy-3-methyglutaryl coenzyme A reductase inhibitors. *Arch. Neurol.* **57,** 1439–1443.

NEUROIMAGING OF COGNITIVE IMPAIRMENTS IN VASCULAR DISEASE

Carol Di Perri,*,† Turi O. Dalaker,*,§ Mona K. Beyer,‡,§ and Robert Zivadinov*,¶

*Department of Neurology, Buffalo Neuroimaging Analysis Center, State University of New York at Buffalo, Buffalo, New York 14203, USA

†Department of Neuroradiology, IRCCS, C. Mondino, University of Pavia, Pavia, Italy

‡Department of Radiology, Stavanger University Hospital, Stavanger, Norway

§The Norwegian Centre for Movement Disorders, Stavanger University Hospital, Stavanger, Norway

¶Department of Neurology, The Jacobs Neurological Institute, State University of New York at Buffalo, Buffalo, New York 14203, USA

Cerebrovascular disease (CVD) is increasingly recognized as a common cause of cognitive impairment and dementia, alone or in conjunction with other neurodegenerative diseases, primarily Alzheimer's disease. The term vascular cognitive impairment (VCI) has been proposed as an umbrella term to recognize the broad spectrum of cognitive changes associated with vascular pathology, mainly characterized by executive impairment together with particular noncognitive features.

DOI: 10.1016/S0074-7742(09)00404-8

0074-7742/09 $35.00

Magnetic resonance imaging (MRI) has become the most important neuroimaging methodology for assessing and monitoring the pathological changes involved in the onset and progression of vascular neurological diseases. Conventional MRI sequences are usually not sufficient to provide full details about the different degrees of brain damage underlying vascular disorders. However, newer and more advanced nonconventional MRI techniques have the capacity to detect invisible brain damage that would otherwise not be detected. This review outlines some of the most important neuroimaging techniques commonly used in the diagnosis, assessment, and monitoring of cognitive impairment in vascular diseases. More detailed applications of these techniques, specifically for VCI, are discussed.

I. Introduction

A. Structural Brain Imaging

Although computed tomography (CT) is still important in many clinical settings, magnetic resonance imaging (MRI) is becoming the imaging method of choice, offering better tissue specificity and sensitivity with respect to relative pathology. Therefore, this section—which discusses structural imaging—will refer only to MRI studies.

1. *Conventional MRI Techniques*

T2-weighted imaging (WI) is highly sensitive in detecting vascular lesions. Upon gliotic transformation, that is, tissue scarring, these appear as local hyperintensities of T2 prolongation in the white matter (WM). There are different and various techniques for identifying T2 hyperintense lesions; among them, the most often recommended are conventional spin echo (CSE), turbo spin echo (TSE), and fluid-attenuated inversion recovery (FLAIR) (Zivadinov and Bakshi, 2004). FLAIR is a MRI sequence for optimized detection of brain parenchymal pathologic water content. It is particularly helpful in the evaluation of periventricular and cortical/juxtacortical lesions, where cerebrospinal fluid (CSF) may mask the visualization of these lesions on T2-WI (Zivadinov and Bakshi, 2004; Zivadinov and Cox, 2007). In the last decade, continuous technical improvements in MRI hardware and software have led to the development of new pulse sequences that are more efficient and sensitive. Among them, TSE or FSE, proton density (PD) and fast-FLAIR have already demonstrated their usefulness in a wide variety of neurologic diseases. Better lesion-to-CSF contrast is achieved with PD because of the relatively lower signal intensity of CSF on this sequence and improved lesion-to-tissue contrast. TSE showed greater sensitivity than CSE in detecting areas of T2 prolongation.

On noncontrast T1-WI, most ischemic lesions are iso- or hypointense in WM. T1-hypointense lesions are an important MRI metric of neurodegeneration in patients with different neurological disorders. They represent a more advanced pathological substrate of the lesions—mainly axonal loss, Wallerian degeneration, and gliotic changes (Zivadinov and Bakshi, 2004; Zivadinov and Cox, 2007).

Even though these techniques are highly sensitive to tissue damage, they are usually not capable of discerning among the possible different pathological origins of the damage (demyelinating, vascular, tumoral). Therefore, both in clinical and research practice, they have to be supplemented with the clinical history of the subject and, most of the time, with other newer nonconventional MRI techniques.

2. *Nonconventional MRI Techniques*

Magnetization transfer imaging. Magnetization transfer imaging (MTI) is an advanced MRI technique that has been widely used to evaluate characteristics and evolution of damaged tissue and normal appearing brain tissue (NABT). It is based on the interactions and energy exchange between protons that are unbound in a free water pool with those where motion is restricted due to binding with macromolecules (Horsfield *et al.*, 2003). MTI uses an off-resonance radio frequency (RF) pulse to saturate protons in macromolecules and water molecules that are bound to macromolecules and are normally not visible due to their very short T2* relaxation times. During pulse sequence acquisition, the saturated protons may enter the free pool of protons, primarily water, or may transfer their magnetization to free water protons. As a consequence, a decrease in the MRI-visible signal is produced in areas characterized by an abundance of macromolecules affected by magnetization transfer. Tissue damage is usually reflected by a reduction in this exchange of protons and thus a decrease in the magnetization transfer ratio (MTR). Decreases in MTR indicate a reduced capacity of free water to exchange magnetization with the brain tissue matrix with which the water comes into close contact. MTRs can be used to detect changes in the structural status of brain parenchyma that may or may not be visible with conventional MRI techniques. Therefore MTI enables semi-quantitative, reproducible characterization of tissue and pathologic entities which could substantially improve the specificity of MRI (Fig. 1).

Diffusion-weighted imaging. Diffusion-weighted imaging (DWI) and diffusion-tensor imaging (DTI) are advanced MRI techniques that allow the measurement of tissue water diffusion-related motion and, as a consequence, provide information about orientation of water motion within the tissue (Rovaris *et al.*, 2005). These techniques take advantage of the direction-dependent mobility of protons which is mainly due to the underlying structure of the brain tissue. The free, random diffusion is restricted by cellular boundaries in different directions to a varying degree, for example, diffusion along a myelinated nerve fiber is much easier than the diffusion perpendicular to it. Similarly, the apparent diffusion coefficient (ADC)

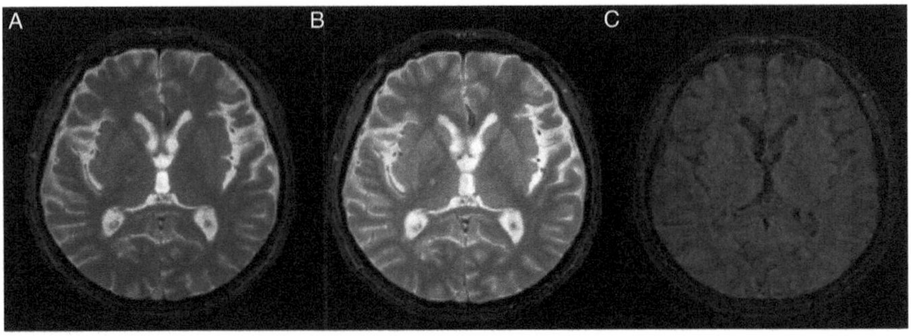

FIG. 1. Magnetization transfer imaging: sample scan with (A) and without (B) magnetic transfer pulse. A relative magnetization transfer ratio map (C) is generated.

is lower in highly structured tissues such as WM and gray matter (GM) than in pure water. ADC values depend on the orientation of the tissue relative to the measurement. Conventionally, the average ADC is calculated from three (DWI) or more than three (DTI) linearly independent directions that provide information on the overall diffusivity in the tissue. Pathological processes that modify tissue organization can cause abnormal water motion, thereby altering ADC values. Ischemic events can alter diffusion parameters of brain tissue *in vivo*, resulting in a decrease of water diffusivity measurable with different DWI and DTI indices. These measures include mean diffusivity (MD), fractional anisotropy (FA), entropy and subsequently tractography outcomes (Fig. 2). It is important to emphasize that, in most instances, very acute stages of ischemic events can be detected through this technique within the first hour of onset. On the other hand, such changes are often not recognized on other MRI sequences. The increased sensitivity of DWI in detecting acute ischemia is thought to be the result of the water shift restricting motion of water protons intracellularly (cytotoxic edema), whereas conventional T2-WI shows signal alteration mostly as a result of vasogenic edema. However, the exact origin of the decreased ADC measurements observed in early ischemia has not been fully established. The diminished ADC value also could be a result of decreased temperature in the malperfused tissues or a combination of all these factors (Busto *et al.*, 1987; Crain *et al.*, 1991).

Regardless of the cause, decreased apparent ADC is a very sensitive indicator of an early ischemic brain event at a stage when ischemic tissue remains potentially salvageable (Wolpert *et al.*, 1993).

Atrophy measures. The measurement of brain atrophy is of growing clinical relevance as a biomarker of the disease process, especially when correlated with other anatomical and functional information from conventional and non-conventional MRI. Various tools are available for semiautomatic evaluation of brain volume, longitudinal studies and intra-/intergroup cross-sectional analysis,

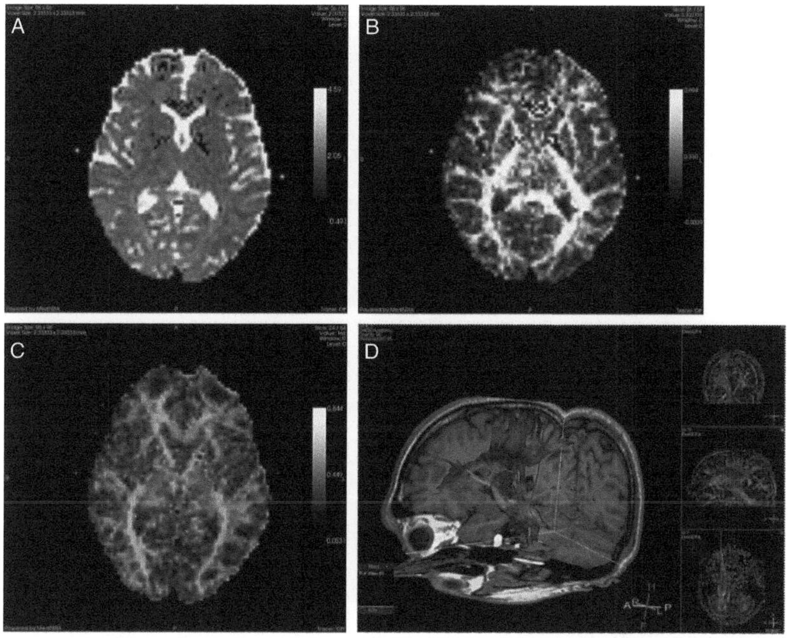

FIG. 2. Diffusion tensor imaging: Mean diffusivity (A), fractional anisotropy (B), and color-fractional anisotropy (C) maps are presented. The maps are defined as functions of the eigenvalues (i.e., the invariants of the diffusion tensor at each voxel); in the RGB (red-green-blue) FA map, the corresponding eigenvectors are also displayed through color coding. A streamline tractographic reconstruction (D) is displayed; fibers are reconstructed according to a linear algorithm following the direction of maximum FA until a threshold value is exceeded. (See Color Insert.)

such as SIENA and SIENAX of the FSL library (http://www.fmrib.ox.ac.uk/fsl/), FreeSurfer (http://surfer.nmr.mgh.harvard.edu/), and SPM (http://www.fmrib.ox.ac.uk/fsl/,http://www.fil.ion.ucl.ac.uk/spm/) (Fig. 3). Voxel-based analysis methods are of growing interest, giving the opportunity to perform local atrophy measurements and possibly displaying a stronger correlation with functional estimates. A high resolution 3D T1-weighted scan is necessary for the analysis and patient age, disease duration, total brain volume or any other presumably significant parameters are typically taken into account as covariates for the statistical analysis. A large subject cohort is indeed necessary for gaining sufficient statistical power, especially considering the sophisticated geometric preprocessing necessary for group analysis.

Atrophy is now proposed to be included as a secondary endpoint in longitudinal trials aimed at monitoring disease progression and therapy efficacy, as complementary information to conventional MRI in various diseases such as multiple sclerosis (MS) and Alzheimer's disease (AD) (Whitwell, 2008).

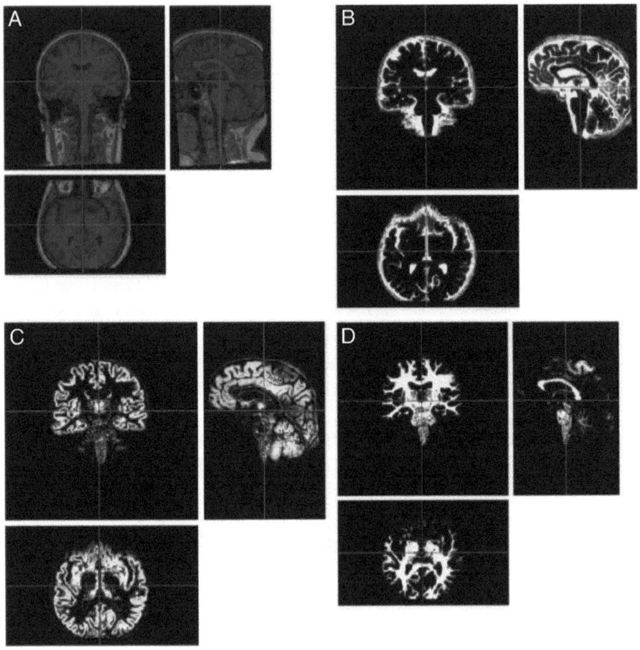

FIG. 3. Brain segmentation with statistical parametric mapping (SPM) software: (A) 3D-T1 image, (B) cerebrospinal fluid, (C) gray matter, and (D) white matter segmentation maps are displayed.

B. Functional Imaging

1. *Perfusion-Weighted Imaging*

Perfusion-weighted imaging (PWI) techniques are sensitive to microscopic levels of blood flow, defined as the amount of blood flowing into a voxel in a certain temporal span and measured in ml blood/min/gr tissue. This index may be altered in disease conditions and accurate monitoring is clinically relevant, improving our understanding of the pathophysiology of the disease.

With respect to other techniques that noninvasively investigate cerebral blood flow, such as positron emission tomography (PET), single photon emission computed tomography (SPECT), and xenon-enhanced computed tomography (Xe-CT), PWI shows some advantages by not using ionizing radiations, allowing a better spatial resolution and a better signal-to-noise-ratio (SNR). In addition, PWI is typically acquired over a short time, needing only a few extra minutes after a conventional MRI exam.

Contrast-enhanced relative cerebral blood volume (rCBV) is the most used perfusion imaging technique for parameter mapping. In this and other techniques, both the ready availability and the T2* susceptibility effects of gadolinium, rather than the T1 shortening effects, make it a suitable agent for use in PWI. As long as the bolus transit through the tissue takes only a few seconds, high temporal resolution imaging is required to obtain sequential images during the wash-in and wash-out of the contrast material and for processing and calculation of hemodynamic maps (including mean transit time (MTT), time to peak (TTP), time of arrival (T0), negative integral (N1). An important neuroradiological indication for MRI is the evaluation of incipient or acute stroke via PWI and DWI. DWI can demonstrate the central effect of a stroke on the brain, whereas PWI visualizes the larger second ring delineating blood flow and blood volume. Echo planar and, potentially, echo volume techniques (such as PRESTO) together with appropriate computing power offer real time images of dynamic variations in water characteristics reflecting perfusion, diffusion, oxygenation, and flow; qualitative and in some instances quantitative maps of regional organ perfusion can thus be obtained (Fig. 4).

2. *Nuclear Medical Imaging*

The general principle behind nuclear medicine imaging is the following: a radioactive biologically active substance is chosen in such a way that its spatial and temporal distribution in the body reflects a particular body function or

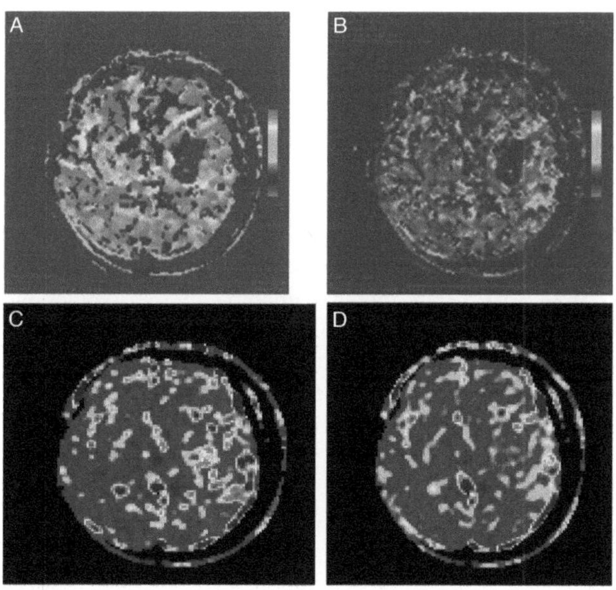

FIG. 4. Perfusion-weighted imaging (PWI): Time to peak (A), mean transit time (B), relative cerebral blood flow (C), and relative cerebral blood volume (D) maps are displayed. These and other possible maps that characterize entity and timing of tissue perfusivity are powerful and sensitive tools for summarizing the dynamic changes conveyed by PWI scans. (See Color Insert.)

metabolism. To study the distribution without disturbing body function, only traces of the substance are administered to the patient.

The radionucleides (radioactive tracers) are unstable nuclei that decay by emission of gamma rays or positrons (followed by annihilation gamma rays). Distribution of the radioactive tracer is inferred from the detected gamma rays, trough instruments known as gamma camera, and mapped as a function of time and/or space. The most often used radionuclides are Tc-99 m in "single photon" imaging and F-18 in "positron" imaging (Saha, 2006).

Currently used nuclear neuroimaging techniques are SPECT and PET. They are both variations on the principle of gamma camera detection of emitted γ-rays. Tomographic reconstruction is used to construct three-dimensional images. Metabolic activity measures, receptor-binding studies, regional blood flow, and drug treatment response can be performed by using radiolabeled ligands. Some applications of PET include evaluation of stroke (blood clot or bleeding in the brain (Lovblad *et al.*, 1997; Sobesky *et al.*, 2005), study of dementia imaging (Silverman, 2004) and evaluation of brain tumors (Tian *et al.*, 2004).

The drawback of both SPECT and PET is poor spatial resolution. In some cases, PET may be more sensitive than SPECT, but PET scanners and tracers are much more costly than SPECT scanners and are often only available in the largest medical centers. A way of overcoming this limit, and thereby improving diagnostic accuracy, is to combine the SPECT and PET scanners with a multidetector computer tomography (MDCT) scanner. SPECT/CT and PET/CT combine a nuclear image with high-resolution structural CT images for accurate localization of lesions and physiological processes (Saha, 2006). Recently, combinations with MRI scanners have also been introduced to the radiological market.

Nuclear medicine imaging is an effective diagnostic tool in the early diagnosis, treatment, and prevention of numerous medical conditions. The main advantage of these techniques is their ability to show molecular function of pathological organs. This allows diagnosing of certain diseases and various medical conditions much sooner than with other medical imaging methodologies which provide mainly structural information about an organ or body part.

C. Metabolic Imaging

1. *Magnetic Resonance Spectroscopy*

In vivo magnetic resonance spectroscopy (MRS) offers information about the biochemistry of a selected brain tissue volume. This can represent surrogate markers for the pathology underlying the pathological process of different neurological diseases (Narayanan *et al.*, 2006).

MRS is a quantitative method of detecting nonaqueous proton signals that correspond to certain biological molecules in the brain. In most conventional and nonconventional MRI techniques, the MR signal originates from water protons and detailed images of the brain are created based on the different water content in various brain tissues. The hydrogen nuclei (protons) that are not in water (nonaqueous protons) are distributed throughout the hundreds of biologically significant molecules of the living brain. These molecular radio-signals are not visible on standard MR images because they are overwhelmed by the high signal derived from the aqueous protons. Each nonaqueous molecule has a unique radio-frequency specific to the chemical environment itself and distinct from the frequency of water protons. The amplitude or strength of these radio-frequencies is dependent on the concentration of the corresponding molecule in the volume of interest, each at its own radio-frequency position.

The most commonly used MRS technique we refer to, proton (1H) spectroscopy, measures the relative amplitude of the corresponding peak for each of several biological molecules by suppressing the signal from aqueous protons (Fig. 5). The best available and most important cerebral components that have been studied are *N*-acetyl aspartate (NAA), choline (Cho), creatinine(Cr)/phosphocreatinine (PCr), myoinositol and lactate. NAA is considered an *in vivo* marker of neuronal integrity and is often measured relative to Cr (which is thought to be unaffected by neurodegeneration). The NAA peak in an MR spectrum is a putative marker of neuronal and axonal integrity, and the Cho peak is largely a marker of cell-membrane turnover. Decreases in Cho usually represent inflammation and demyelinating processes, while decreases in NAA are markers of axonal and neuronal injury. On this basis, a reduction in the NAA peak is interpreted as representing neuronal/axonal dysfunction (Narayana, 2005) and an elevated Cho peak represents heightened cell-membrane turnover, as seen mainly in tumors. Therefore, MRS may provide unique information regarding metabolic changes in the CNS, evaluation of the pathogenesis, severity and progression of the neurological disease. Other metabolic peaks of increasing interest in the study of neurologic disorders are the glutamate/glutamine peak, representing a mixture of amino acids and bioamines used throughout the CNS as excitatory and inhibitory neurotransmitters (Vrenken *et al.*, 2005), and the myoinositol peak, representing a sugar-like molecule thought to be a marker of glial proliferation and now recognized for its importance in osmotic regulation of brain tissue (Narayana, 2005).

High-resolution ^{1}H MRS is expected to improve our understanding of the metabolic alterations that accompany ischemia, particularly those related to acid-base balance. ^{1}H MRS can depict an increase in lactate concentration and concomitant decrease in intracellular pH in the hyperacute ischemic stage when conventional MRI signal abnormalities are minimal.

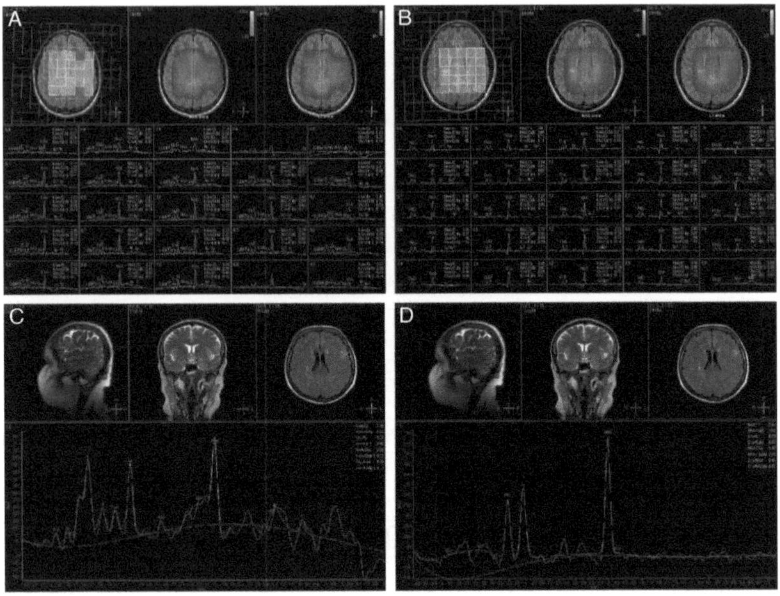

FIG. 5. Multi-voxel sMR: a TE = 25 ms (A) and TE = 144 ms (B) scan are reported with the fitted spectrum and estimated peak ratio reported for each voxel. A mixed Gaussian-Lorenztian peak shape was chosen for the fit as the most general. Single-voxel sMR: a TE = 25 ms (C) and a TE = 136 ms (D) scan are reported with the corresponding fit overlaid. The higher biochemical specificity of short TE scans is evident at the expense of the less accurate fit due to an irregular baseline and to peak overlap. (See Color Insert.)

II. Vascular Cognitive Impairment

A. Terminology

After AD, CVD is recognized as the second most common cause of acquired cognitive impairment and dementia, alone or in conjunction with other neurodegenerative disorders (O'Brien, 2006).

Progress in the concept of vascular cognitive impairment (VCI) has been seriously limited by difficulties in finding agreement with respect to terminology. O'Brien *et al.* (2003) have proposed using VCI as an umbrella term, including all aspects and degrees of cognitive impairment resulting from vascular brain damage. Conversely, some authors favor regarding vascular dementia (VaD) and VCI as two different entities (Roman *et al.*, 2004). Although several definitions exist, both DSM-IV and ICD-10 require the presence of memory impairment as an absolute requirement, in addition to one or more other cognitive domains being affected. This has been largely criticized, since the presence of memory impairment is known to be more prevalent in neurodegenerative disease, such as AD, than in CVD. The use of a term of dementia that does not require memory impairment and instead focuses on cognitive symptoms such as executive deficits (known to be affected more in CVD) would possibly solve this dilemma. For the purpose of this review, the term VCI will be considered to refer to all forms of mild to severe cognitive impairment associated with and presumed to be caused by CVD (O'Brien, 2006; O'Brien *et al.*, 2003).

B. Classification

The proposed classification by O'Brien *et al.* (O'Brien, 2006; O'Brien *et al.*, 2003) distinguishes between sporadic and hereditary disorders that represent the two main categories. The sporadic VCI is further divided into several clinical subtypes: post-stroke dementia (PSD), VaD, mixed AD and VaD (mixed dementia) and, finally, vascular mild cognitive impairment (VaMCI). The term VaD is again subclassified into multi-infarct dementia, subcortical ischemic vascular dementia (SIVD), strategic-infarct dementia, hypoperfusion dementia, hemorrhagic dementia, and dementia caused by specific arteriopathies.

1. *Classification of Sporadic Vascular Cognitive Impairment*

Sporadic VCI is a clinical-pathological entity which includes several different sporadic vascular disorders which are not yet fully defined, whose main feature is their contribution to cognitive impairment.

Post-stroke dementia. PSD includes all types of dementia that happen after stroke, irrespective of their cause (Leys *et al.*, 2005). Their prevalence is likely to increase in the future, because of the decline in mortality after stroke (Rothwell *et al.*, 2004) and the aging population. Community-based studies suggest that having a stroke doubles the risk of dementia (Leys *et al.*, 2005). PSD might be the result of vascular lesions, AD, WM changes or a combination of these. The causes of PSD differ among studies in relation to patient age, time after stroke, ethnicity and criteria used. The risk factors are in general the same as for most vascular disease (arterial hypertension, history of high cholesterol, diabetes, forms of heart disease). Stroke-related variables associated with an increased risk of PSD are stroke severity, cause, location, and recurrence (Leys *et al.*, 2005). Most studies have found that a more severe clinical deficit at onset is associated with a higher risk of PDS (Tatemichi *et al.*, 1992, 1990). The risk of PSD and its severity are not influenced by the type of stroke (ischemic or hemorrhagic) (Barba *et al.*, 2000; Linden *et al.*, 2004) (Fig. 6).

Vascular dementia. Multi-infarct dementia, also known as cortical vascular dementia, is the traditional term used for VaD and the consequence of multiple large cortical infarcts (Hachinski *et al.*, 1974). It is clinically characterized by a

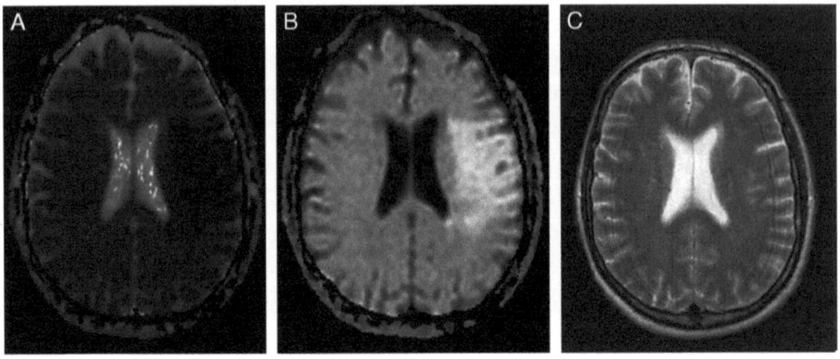

FIG. 6. Acute stage of left medium cerebral artery ischemic stroke: as shown on the images, the ischemic acute stage is visible on DWI-weighted images earlier than on conventional MRI: (A) ADC (B) eADC, and (C) conventional T2-weighted MRI scan.

relatively abrupt onset (days to weeks), a stepwise deterioration and a fluctuating course. It is related predominantly to large vessel disease and cardiac emboli events, being characterized by cortical-subcortical, arterial, territorial, and distal field infarcts (watershed areas).

Subcortical ischemic vascular dementia. SIVD is attributed to small-vessel disease and it is characterized by lacunar infarcts, ischemic WM lesions, and incomplete ischemic injury (Decarli, 2004; Wallin *et al.*, 2003). The clinical onset is variable (Babikian and Ropper, 1987) and the course is gradual, with or without acute deficits. The early cognitive syndrome of SIVD is characterized by a dysexecutive syndrome with slowed information processing. The dysexecutive syndrome includes impairment in goal formulation, organizing, planning, executing, and abstraction (Cummings, 1994). Memory loss is usually milder than in AD. In this subgroup of VaD, the onset is often more insidious and temporal relations between cognitive syndrome, brain imaging findings and vascular disease may not be clear (O'Brien *et al.*, 2003).

Hemorrhagic dementia. Hemorrhagic dementia is usually due to uncontrolled hypertension or vascular malformations, subarachnoid hemorrhage and intracerebral hemorrhage, with or without vasospasm. Due to the extent of the hemorrhage, patients often show additional neurological impairments such as sensorimotor changes or gait impairment. One of the most common causes of hemorrhagic dementia is subdural hemorrhage. Although it is primarily post-traumatic, considering it a vascular injury makes subdural hematoma a common and insidious cause of VCI in clinical practice (Starkstein *et al.*, 2005).

Much more common than epidural hemorrhages, subdural hemorrhages generally result from shearing injuries due to various rotational or linear forces (Maxeiner and Wolff, 2007). The disease is commonly seen in the elderly and in alcoholics, who have evidence of brain atrophy. Cerebral atrophy increases the length the bridging veins have to traverse between the two meningeal layers (dura and arachnoidea), hence increasing the likelihood of shearing forces causing a tear. It is also more common in patients on anticoagulants, who can have a subdural hematoma with a minor injury. Chronic subdural bleeds can develop over a period of days to weeks, often after minor head trauma, though such a cause is not identifiable in 50% of patients. They may not be discovered until they present clinically months or years after a head injury and they are rather common in the elderly (Kushner, 1998) (Fig. 7).

Strategic-infarct dementia. It is characterized by focal, often small, ischemic lesions involving specific sites critical for specific higher cortical functions. Cortical sites more often affected are the hippocampal formation, angular gyrus and cingulate gyrus and thalamic infarcts (bilateral or monolateral).

Dementia caused by specific arteriopathies. Arteriopathies such as cerebral vasculitis, that is, acute or chronic inflammatory changes of small, medium, and large arteries or veins, can be cause of multiple ischemic lesions and lead to VCI. They are often accompanied by systemic signs of fever, malaise, and weight loss.

Some rare arteriopathies such as inflammatory arteriopathy (e.g., polyarteritis nodosa, temporal arteritis) and noninflammatory arteriopathy (e.g., moyamoya disease, fibromuscular dysplasia) can cause multiple infarcts and lead to vascular dementia as well. In cerebral amyloid angiopathy-associated vasculopathy, aneurysm formation and stenosis in the leptomeningeal and cortical vessels cause damage to the subcortical WM. In hereditary cystatin-C amyloid angiopathy, patients have recurrent cerebral hemorrhages before the age of 40 that can lead to dementia.

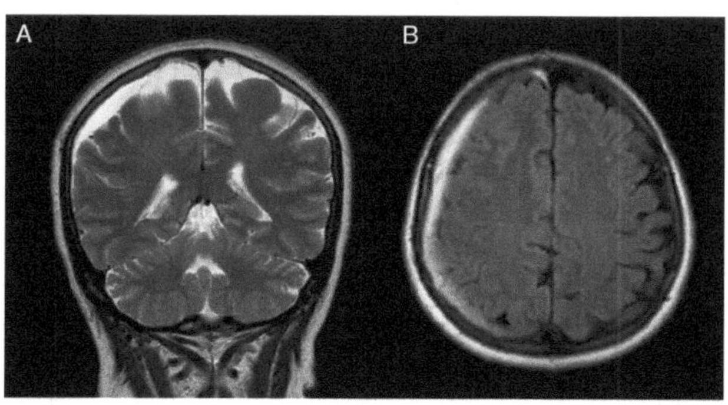

FIG. 7. Coronal and axial images of subdural hemorrhage on conventional MRI T2-weighted sequence (A and B).

Hypoperfusion dementia. Hypoperfusion due to large vessel or cardiac disease can affect the watershed areas of the brain and lead to vascular dementia. A correlation between systolic blood pressure reduction and cognitive decline in women, which was not accounted for by other factors, has been recently demonstrated (Zhu *et al.*, 1998). Baseline blood pressure was not found to be significantly correlated to cognitive decline with initial good cognition (Zhu *et al.*, 1998). It has also been suggested (Skoog *et al.*, 1998) that age-related changes in brain structure may contribute to the decrease in blood pressure in the very elderly, and that low blood pressure in dementia disorders is mainly a secondary phenomenon. Some researchers, therefore, have speculated that blood pressure reduction might be an early change in the dementing process, but further studies are necessary to clarify this aspect.

Mixed AD and VaD (mixed dementia). Mixed dementia has for a long time been underestimated as a common cause of dementia, particularly in the elderly. In fact, its diagnosis is still mostly a challenge. It has been shown that different vascular factors, including arterial hypertension and stroke, increase the risk of AD. Frequently, CVD coexists with AD (Decarli, 2004; de la Torre, 2004). This overlap is increasingly important in the elderly population since CVD is a potentially preventable and treatable cause of dementia. The actual relationship between AD and vascular pathology is still unclear. They may represent two independent co-occurring pathologies which happen to share risk factors (Decarli, 2004). However, accumulating evidence shows that vascular changes may actually stimulate or exacerbate the formation of AD pathology. For example, vascular disease might determine vessel wall thickening and reduce the efficiency of the perivascular drainage system, leading to a reduced rate of elimination of amyloid (Nicoll *et al.*, 2004) through the drainage system. Some of the most challenging clinical scenarios include VaD patients with an insidious onset or slow progression and AD patients with evidence of vascular lesions upon brain imaging. Given that reliable biological markers for recognizing AD and discerning it from mixed dementia do not yet exist, neuroimaging plays an important role as a potential marker. In fact, besides early episodic memory impairment and low concentrations of CSF peptides with high tau protein concentrations, early and significant medial temporal lobe atrophy on MRI and bilateral parietal hypoperfusion on SPECT are strongly suggestive of AD (Erkinjuntti, 2007).

Vascular mild cognitive impairment. VaMCI is a low degree of cognitive impairment resulting from CVD which does not meet the criteria for dementia. This entity has been relatively underestimated and understudied compared to the nonvascular pre-Alzherimer's (MCI) (Petersen *et al.*, 1999). However, recent studies have shown a higher percentage of VaMCI than VaD (Rockwood *et al.*, 2000) and a high tendency for converting into dementia in a relatively short time (Wentzel *et al.*, 2001).

2. *Hereditary Vascular Cognitive Impairment*

Understanding of the pathogenesis of vascular disease has improved greatly in recent years, and many molecular genetic conditions related to this symptom have been found. Among them, there is strong interest in cerebral autosomal dominant arteriopathy with subcortical infarcts and leucoencephalopathy (CADASIL).

CADASIL is an inherited CVD due to mutations of the Notch3 gene at the chromosome locus 19p13. The clinical spectrum includes recurrent ischemic episodes, cognitive deficits, migraine and psychiatric disorders. The histopathological hallmark of CADASIL is accumulation of electron dense granules (GOM) in the media of arterioles. CADASIL was initially thought to be a rare disorder, but increasing numbers of families have been identified; therefore, it is likely that CADASIL is still largely underdiagnosed (Federico *et al.*, 2005). The course of the disease is very heterogeneous, even in the same family; some patients remain asymptomatic until their 70s, whereas others are severely affected from the age of 50 onward (Fig. 8).

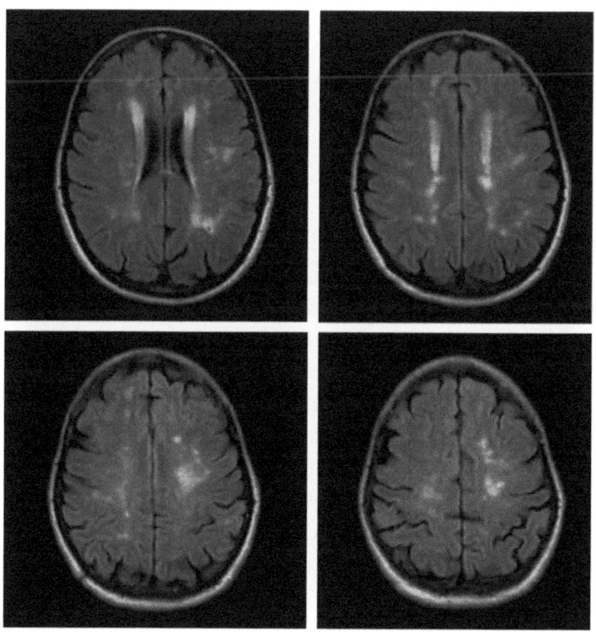

FIG. 8. Axial FLAIR sequence displays sample images of CADASIL MRI pattern of conventional MRI in a 32 y.o. patient.

C. Histopathology of VCI

Vascular disease produces either focal or diffuse effects on the brain and may cause cognitive decline. Focal CVD occurs mainly secondarily to thrombotic or embolic vascular occlusions. Common areas of the brain associated with cognitive decline are the WM of the cerebral hemispheres and the deep GM nuclei, especially the striatum and the thalamus. Hypertension is the major cause of diffuse disease and, in many patients, both focal and diffuse disease forms are observed together. The five most common pathological findings of VaD are white matter hyperintensities (WMH), macroscopic, lacunar and/or microscopic infarcts and microbleeds.

1. *White Matter Hyperintensities*

WMH, as revealed by MRI-histopathological studies, are associated with evidence of hypoxic tissue injury (Fernando *et al.*, 2006), with reduced capillary density not only in the lesions but also in normal-appearing WM (Brown *et al.*, 2007). However, smooth periventricular lesions directly adjacent to the ventricles of high T2 signal are considered by some researchers to be of nonischemic origin and consist of areas of demyelination, subependymal gliosis and discontinuity of ependymal lining (Fazekas *et al.*, 1993).

The most important risk factor for WMH is age, but other known risk factors are hypertension and decreased peak expiratory flow, elevated levels of glycated hemoglobin (Murray *et al.*, 2005), type 2 diabetes (Korf *et al.*, 2006) and cigarette smoking (Liao *et al.*, 1997).

2. *Macroscopic Infarcts*

Infarctions involving areas of major cerebral arteries may result in large territorial lesions with a variable amount of tissue loss. The infarcts develop in typical stages of gliosis with increased water content. The intensity of gliotic scarring is used in pathological studies to judge the degree and age of the infarction (Jellinger, 2007). Hypertension, diabetes, hypercolesterolemia, smoking, and heart disease are the most common risk factors for infarcts.

Multiple cortical infarcts are believed to be the neuropathological cause of multi-infarct dementia (Hachinski *et al.*, 1974). In multi-infarct dementia, the combined effects of different infarcts produce cognitive impairment by affecting neural networks. The cognitive impact of large complete infarcts depends on localization and the extent of the ischemic lesion.

3. *Lacunar Infarcts*

Lacunes (small infarcts in WM and deep GM structures with diameters ranging from 3 to 15 mm) (Jellinger, 2007), represent different stages of ischemic injury exhibiting loss of axons and myelin together with reactive changes (Ishii *et al.*, 1986). In pathological studies, they have been associated with cognitive decline independent of other dementia pathology such as AD (Gold *et al.*, 2005). Some studies have reported that subcortical lacunes and multiple widespread infarcts are the most common morphologic substrates of VaD (Jellinger, 2007).

Lacunes in SIVD are typically located in the caudate, globus pallidus, thalamus, internal capsule, corona radiata, and in frontal WM (O'Brien *et al.*, 2003). Lacunar state is a condition in which numerous lacunae are present, which indicates widespread severe small-vessel disease. It is due to small-vessel disease affecting all the small vessels of the brain. The amount of lacunes has been found to be a significant predictor of cognitive status in the elderly and of executive dysfunction in patients 3 months after an ischemic stroke (van der Flier *et al.*, 2005; Vataja *et al.*, 2003).

4. *Microscopic Infarcts*

Recently, there has been increasing focus on the contribution of cortical microinfarcts (not visualized by current available neuroimaging techniques) to cognitive decline both in normal aging (Kovari *et al.*, 2004), VaD and mixed dementia (Gold *et al.*, 2007a). Cortical microinfarcts might be as strong a predictor of cognitive decline as is the presence of neurofibrillary tangles in AD (Gold *et al.*, 2007b).

5. *Microbleeds*

Microbleeds (MBs), characterized histopathologically by the presence of hemosiderin around small vessels (Fazekas *et al.*, 1999), display themselves in several ways, mainly as hypertensive arteriopathy or amyloid angiopathy (Fazekas *et al.*, 1999). MBs are often detected in demented individuals, such as patients with AD (Hanyu *et al.*, 2003), CADASIL (Dichgans *et al.*, 2002), subcortical vascular dementia (Won Seo *et al.*, 2007) or memory loss (Cordonnier *et al.*, 2006). Cordonnier *et al.* (2006) studied the prevalence of MBs in a large cohort of patients (772 patients) attending a memory clinic. They found that 65% of patients with vascular dementia exhibited MBs, vs 18% of AD patients, 20% of mild cognitive impairment patients, and 10% of patients with subjective complaints. The presence of MBs was associated with age, WMH, lacunar infarcts, and infarcts. This finding of a relatively high proportion of MBs in AD and mild cognitive impairment also provides further evidence for the involvement of vascular factors in neurodegenerative diseases such as AD.

D. Diagnostic Neuroimaging Criteria in VCI

Conventional imaging plays a central role in the diagnostic work-up in most vascular disorders. It is suggested that MRI should be incorporated as an essential part of clinical trials involving VCI (O'Brien *et al.*, 2003; Schmidt *et al.*, 2007), but the pulse sequences, techniques and clinical outcome to be used remain a matter of debate (Enzinger *et al.*, 2005; Schmidtke and Hull, 2005).

Of the available diagnostic criteria for VaD, only two of them, the NINDS-AIREN (Roman *et al.*, 1993) and the State of California AD Diagnostic and Treatment Centers (ADDTC) criteria (Chui *et al.*, 1992), require radiological findings of CVD in the form of infarcts or WMH.

E. Therapeutic Strategies

Therapeutic approaches can be divided into primary prevention (preventing the occurrence of CVD), secondary prevention (Fig. 9) (preventing the exacerbation of CVD), and symptomatic treatments (O'Brien, 2006).

Much is known about primary prevention, which mainly consists of preventing the vascular disease leading to VCI by reducing the above-mentioned risk factors (mainly hypertension, cigarette smoking, hyperlipidemia, etc.). However, the degree to which primary prevention can affect nonstroke-related damage and, in particular the development of WMH, is still mostly unclear. Secondary prevention consists mostly of preventing recurrent vascular episodes, mainly strokes, by the use of antiplatelet agents, carotid endarterectomy, anticoagulants for atrial fibrillation and decreasing blood pressure (O'Brien, 2006).

Regarding symptomatic treatments, several different trials have been undertaken, although they have been largely criticized, mainly because they have focused more on AD than VCI (the AD group is much bigger, and the effect on society would thus be more significant). One single study in VaD showed that aspirin yielded better cognitive outcomes for treated as opposed to untreated patients (Meyer *et al.*, 1989). Many studies suggest that cholinesterase inhibitors may prove beneficial in the treatment of VaD, both in cognitive and behavioral symptoms, as they already have in the management of AD (Erkinjuntti *et al.*, 2002; Pratt and Perdomo, 2002).

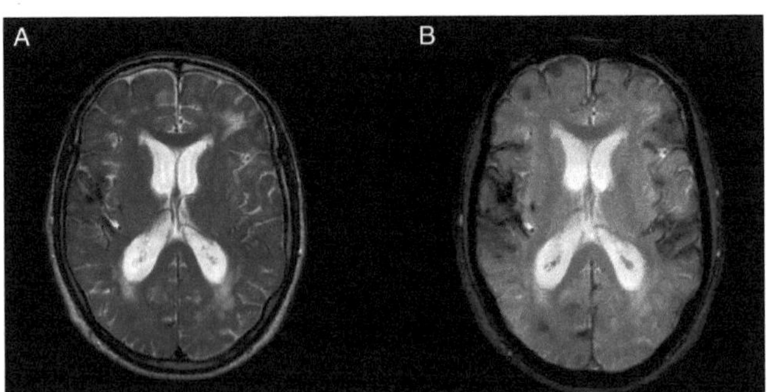

Fig. 9. T2w (A) and GRE axial images (B) in a case of amyloid angiopathy.

With respect to neuroimaging, such treatment effects would be especially interesting considering primary prevention strategies. The various neuroimaging techniques have, as shown below, been able to detect both established CVD as well as preclinical brain damage in healthy individuals possibly at risk. In this respect, neuroimaging may serve as an important tool of the future in attempts to reduce the prevalence of VCI.

III. Updates on Neuroimaging Techniques in VCI

A. Conventional Structural MRI

Conventional MRI techniques play an important role both for diagnosis and prognosis in VCI patients. Our knowledge of both incidental and disease-related findings has greatly increased during the last decades and allowed a better understanding of the pathological basis of VCI. However, VCI shows no pathognomonic imaging features. Infarct location often does not correlate with the cognitive profile, and neuroimaging can establish the chronology of lesions only with certain limits and cannot offer information about the relative contribution of neurodegenerative versus ischemic processes to the clinical presentation.

1. *White Matter Hyperintensities*

Apart from providing indications of ischemic lesions, MRI has, due to its widespread availability, resulted in increased recognition that WMH on T2-WI are common incidental findings in the brains of otherwise clinically healthy individuals (Fig. 10) (Enzinger *et al.*, 2006; Fazekas, 1989; Wen and Sachdev, 2004). In fact, epidemiological imaging studies show that WMH are extremely

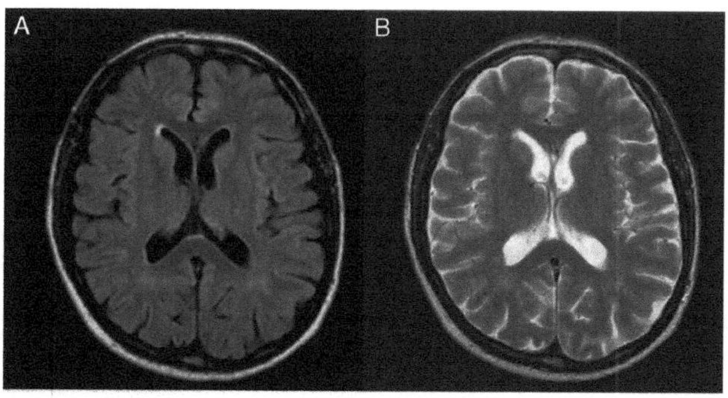

FIG. 10. Normal findings in healthy individuals on conventional MRI: (A) axial FLAIR image, (B) axial T2-weighted image.

common in the elderly (de Leeuw *et al.*, 2001; Liao *et al.*, 1997; Wen and Sachdev, 2004). In a population-based study, de Leeuw *et al.* (2001) showed that, of 1077 subjects between 60 and 90 years of age, only 5% had no WMH detectable on MRI. Other studies have shown up to 90% prevalence of WMH (Sachdev *et al.*, 2005). Most studies show that women have more WMH changes than men (de Leeuw *et al.*, 2001; Sachdev *et al.*, 2007; van den Heuvel *et al.*, 2004), while some others show contradictory findings (Wen and Sachdev, 2004).

Accumulation of WMH around the ventricles, also called leukoaraiosis, appears as a bright signal on T2-WI, indicating pathology involving increased water content and gliotic scarring in the WM of the brain. The FLAIR sequence is commonly used for detection of these WMH. Typical WMH in subcortical ischemic disease include extensive periventricular and deep WM signal abnormalities, usually located in the genu and anterior limb of the internal capsule, anterior corona radiata and anterior centrum semiovale (O'Brien *et al.*, 2003).

In nondisabled elderly living independently, age-dependent WMH have been recently reported as relating to both global cognitive dysfunction and various specific cognitive performances such as executive function, attention, and speed processing (Verdelho *et al.*, 2007). Despite evidence that age-related WMH influence cognitive status, some studies report that there is no such relationship, suggesting that MRI might be oversensitive when it comes to WMH, and that higher water content in the brain does not necessarily result in loss of function (Schmidtke and Hull, 2005). The association between WMH (especially low grade) and cognition in nondemented individuals is at the present time complex and not fully understood.

In subjects affected by VaD, the NINDS-AIREN criteria (Roman *et al.*, 1993) require that WMH involve at least 25% of total brain WM. In order to be significant, WMH must be diffuse and characterized by irregular periventricular lesions extending into deep WM, sparing areas thought to be protected from perfusion insufficiency (e.g., subcortical U-fibers and external capsule-claustrum-extreme capsule). Studies on the relationship between WMH in VaD and cognitive impairment have led to contradictory results (Cohen *et al.*, 2002; Fernando and Ince, 2004).This can be explained by the fact that the relationship between WMH and dementia might be driven by a threshold effect (Libon *et al.*, 2008; Wright *et al.*, 2008), which implies that a certain amount of WMH is needed to have clinical consequence. Once this threshold is reached, other factors might contribute more to cognitive impairment than WMH volume (Sweet *et al.*, 2003). In VaMCI patients, recent studies have also found an association between WMH volume and deficits on cognitive tests (Debette *et al.*, 2007; Sachdev *et al.*, 2006; Yoshita *et al.*, 2006), especially for executive dysfunction (Bombois *et al.*, 2007). Furthermore, periventricular WMH have been found to predict conversion to VaD and mixed dementia, but not to increase the risk of developing other dementias like AD, DLB, and fronto-temporal dementia (Bombois *et al.*, 2008).

2. *Macroscopic, Lacunar, and Microscopic Infarcts*

Cerebral infarction is focal brain necrosis due to complete and prolonged ischemia that affects all tissue elements, neurons, glia, and vessels. As previously described, they differ from each other with regards to their size and spatial location. The clinical consequences vary as well, depending on the size and location of the infarct, that is, a very small infarct in a core region such as the rolandic area can lead to pronounced deficits with very poor outcome, whereas a larger ischemic event in a not "eloquent" location can be of much less importance, that is, remain clinically silent. On conventional MRI such infarcts are seen as hyperintensities on T2-WI regardless of their cause, and they can be detected as soon as they enter into a subacute stage. For detecting the acute stage of an infarct and, therefore, the therapeutical possibilities of limiting the structural damage and neurological consequences, nonconventional techniques play a superior role.

The amount of lacunes has been found to be a significant predictor of cognitive status in the elderly (van der Flier *et al.*, 2005), in CADASIL disease (Liem *et al.*, 2007), as well as in the executive dysfunction of nondemented patients (Carey *et al.*, 2008) and in post-stroke patients (in the last ones in conjunction with temporal lobe atrophy, WMH volume, education, and Mini-Mental State Examination (MMSE)) (Vataja *et al.*, 2003). However, an earlier study found no association between either volume or localization of lacunes and cognition in SIVD (Fein *et al.*, 2000). The exact impact of lacunes on cognitive decline in SIVD is, therefore, not yet established and fully understood.

3. *Microbleeds*

In patients with lobar intracranial hemorrhage, cognitive decline has been shown to be more severe in subjects with larger numbers of MBs at baseline (Werring *et al.*, 2004). In a recent study, MBs appear to be primarily associated with global cognitive impairment even in subjects with no history of neurological disorder (Cordonnier *et al.*, 2006; Werring *et al.*, 2004).

4. *Hereditary VCI*

In hereditary VCI such as CADASIL, conventional MRI reveals extensive cerebral WM lesions, subcortical infarcts and MBs. CADASIL is characterized by WMH on T2-WI in the subcortical WM and basal ganglia. Two major types of abnormalities were first observed by Skehan *et al.* (1995) and later confirmed by other studies (Lesnik Oberstein *et al.*, 2003). The most striking findings were large confluent patches of high-signal change on T2- and proton density-weighted images present throughout the white matter, especially in the anterior part of the temporal lobes and the periventricular portion of the occipital lobes. Additionally, small linear and punctuate lacunes were detected, present not only in the periventricular white matter but also in the brain stem, basal ganglia, thalamus, external capsule, and corpus callosum (Skehan *et al.*, 1995).

One of the main disabling symptoms of this disease is the cognitive impairment, providing CADASIL as a pure VCI entity. A recent study by Viswanathan *et al.* (2007) showed that, among the lesions observed on conventional MRI in CADASIL, the overall lacunar lesion burden seems to have the most important impact on cognitive function and disability. These findings suggest that preventive strategies to decrease the risk of lacunar lesions as observed on MRI may reduce disease-related impairment in CADASIL. These results suggest that lacunar lesions may also play a key role in disability and cognitive impairment in more common forms of small-vessel disease.

5. *Brain Atrophy*

Analyses estimating brain volumes have been widely used with respect to ischemic pathology. In normal aging, a large epidemiological study found that high WMH volume correlated with GM atrophy (Wen *et al.*, 2006). Medial temporal lobe atrophy (MTA) was recently demonstrated to predict cognitive status after stroke alone (Firbank *et al.*, 2007) or in conjunction with brain infarct, WMH volume, and the subject's educational level (Pohjasvaara *et al.*, 2000).

In CADASIL, atrophy is currently investigated as an additional important characteristic of the disease (O'Sullivan *et al.*, 2007; Peters *et al.*, 2006). Cerebral volumetric assessment might thus be an additional aspect in understanding the pathogenesis behind neuropsychiatric and neurobehavioral changes in ischemic brain diseases. A limitation to keep in mind is that brain atrophy might develop rather late in the course of some diseases and for the time being serves mostly to characterize the natural development of the pathology.

B. Nonconventional Techniques

1. *Diffusion-Weighted Imaging and Diffusion-Tensor Imaging*

Data from healthy individuals and patients with MCI and dementia, independent of their cause, converge on highlighting a positive correlation between cognitive performance and fractional anisotropy, and a negative correlation between cognitive performance and mean diffusivity (Kapeller *et al.*, 2004). Moreover, correlations between diffusional measurements and behavioral scores (e.g., side-alternating tapping speed) have been reported for normal individuals.

Diffusion tensor variants have also been previously shown to yield a better correlation with cognition than conventional MRI measures in VCI. DTI indices have been proved to correlate much more strongly with cognitive function than

WMH volume in patients with ischemic lesions. The correlation was strongest for diffusivity of normal appearing WM (NAWM), and remained significant after controlling for conventional MRI parameters, including brain parenchymal volume and T1 and WMH volumes. Regarding conventional measurements, brain volume predicted cognitive impairment best. Of special interest is the finding that DWI/DTI was able to detect specific pathology in NAWM, demonstrating the technique as a highly sensitive and early predictor of brain damage. This is in line with a pathological study combining postmortem MRI and histology (Brown *et al.*, 2007), and shows the great potential of newer imaging techniques. DWI might thus serve as a promising imaging technique for further knowledge about vascular changes in the brain not visible on traditional MRI scans and for exploring the mechanisms of cognitive dysfunction in these patients.

2. *Magnetization Transfer Imaging*

In many previous studies, WMH were not further evaluated than assessing the total volume of WMH based on T2-WI. As a result some studies show only correlations, if any, between the WMH volume on one hand and measures of cognitive function or risk factors on the other.

Recently, MTI has demonstrated that age-related WMH are quite heterogeneous despite their similar appearance on T2-WI, which is assumed to reflect histological differences in these lesions (Spilt *et al.*, 2006). Thus, by taking into account the heterogeneity of age-related WMH, both in terms of etiology and in terms of the severity of tissue destruction, a better understanding of the causes and consequences of these lesions and other tissue structure abnormalities leading to VCI can be obtained. This might explain previous contradictory findings on the clinical impact of WMH, since these lesions are not uniform.

C. Metabolic Imaging

1. *Magnetic Resonance Spectroscopy*

MRS has been shown to be superior to conventional MRI measurements in predicting cognitive decline after stroke. In a longitudinal study (Ross *et al.*, 2006), reduced frontal WM NAA/Cr and NAA appear to be useful predictors of cognitive decline over 12 months and 3 years in stroke/TIA and healthy aging, respectively. MRS may therefore play a useful role in the early investigation of individuals at increased risk of cognitive decline after stroke/TIA. Individuals identified to be at risk for cognitive decline may potentially benefit from early interventions. These MRS results were better predictors than hippocampal volume, whole brain or WMH volumes. As suggested by the authors, assessment of frontal NAA/Cr ratio might serve as a possible biomarker for identifying patients at risk for cognitive decline after stroke/TIA.

D. Functional Imaging

Contrary to the case in neurodegenerative disorders, functional imaging has played a minor role compared to structural imaging in the investigation of cognitive impairment in CVD. Despite this, its use has been recently growing. It is therefore proposed by Nagata *et al.* (2007) that the old VaD criteria should be revised especially due to new insight brought by nuclear imaging techniques.

1. *Perfusion-Weighted Imaging*

Zimny *et al.* (2007) studied 64 patients with dementia with different degrees of cognitive impairment to assess the relationship between cognitive impairment according to the MMSE and values of CBF, CBV, and MTT obtained in perfusion CT (pCT). The results of this study showed that CBF and CBV calculated with pCT correlate with cognitive impairment in patients with dementia and thus may play a role in monitoring disease progression or therapeutic response (Zimny *et al.*, 2007). This may be a good setting for developing new studies using either pCT or perfusion MRI, the latter losing in spatial resolution but gaining in tissue specificity.

2. *SPECT and PET*

Functional imaging studies in VCI have mainly focused on cerebral metabolism and rCBF using techniques such as 18F-2-fluorine-2-deoxyglucose (FDG)-PET and perfusion SPECT. The pattern of altered brain metabolism is more variable in vascular disorders than in neurodegenerative, reflecting the heterogeneity of the lesions involved. Still, a few studies have found that functional nuclear imaging can distinguish VaD from differential diagnoses such as AD (Kerrouche *et al.*, 2006; Mielke *et al.*, 1994; Nagata *et al.*, 2000) and frontotemporal dementia (Lojkowska *et al.*, 2002). Functional imaging can be used to see the remote effect of different ischemic lesions on higher cerebral functions. In the past few years, radiotracer imaging has been used in many studies to address this issue.

For lacunar infarcts, Clarke *et al.* (1994) found that an isolated infarct in the left anterior nuclei of the thalamus in a patient with clinical amnesia was associated with reduced metabolism in the thalamus, amygdala, and posterior cingulate cortex. These are all regions known to be involved in memory functions. Another study found that not only symptomatic but also silent lacunes were associated with global reduction of brain glucose metabolism (Takahashi *et al.*, 2000).

PET and SPECT have also evaluated the effect of WMH on cerebral metabolism (Clarke *et al.*, 1994; Sultzer *et al.*, 1995; Takahashi *et al.*, 2000). In a FDG-PET study of subjects with no signs of neurological disorders, Takahashi *et al.* (2000) found that an increasing severity of incidental WMH, particularly periventricular lesions, was associated with reduced cerebral metabolism. Furthermore, the periventicular lesions were associated with reduced performance on tests for attention and speed. An FDG-PET study (Sultzer *et al.*, 1995) investigated the effect of various ischemic subcortical lesions on cortical metabolism in patients with VaD. This study, with relatively few subjects and heterogeneity in lesion types, found that the type of subcortical pathology had a different influence on the amount of reduced metabolism in the cortex. In general, reduced metabolism was associated with the presence of WMH and lacunes. In addition, the severity of WM lesions correlated with anxiety, depression and overall severity of neuropsychiatric symptoms. Reed *et al.* (2004) performed a PET study investigating the impact of subcortical ischemic lesions on neuropsychological status in ischemic VaD, taking cerebral glucose metabolism, and both hippocampal and cortical atrophy, into consideration. The results showed that the functional impact of subcortical lesions, both lacunes and WMH, was strongest in the dorsolateral frontal cortex. The MRI pathology correlated with hypometabolism here and also with reduced executive function on neuropsychological tests. Hypometabolism and dysexecutive function was also associated with cortical atrophy, suggesting that this contributed to the clinical and imaging findings.

In SIVD, Yang *et al.* (2002) reported that, rCBF measured by SPECT showed a characteristic pattern with reduced metabolism in deep GM structures, superior temporal, and ventral subcallosal gyri. This pattern correlated with cognitive dysfunction. Taken together, the above studies suggest that vascular pathology may affect cerebral metabolism in a way that contributes to the cognitive dysfunction observed in VCI.

E. Correlations Between Imaging Findings and Pathology

Regarding ischemic changes in WM, lesion size is found to correlate with severity of pathologically confirmed tissue damage (Fazekas *et al.*, 1993). A recent combined neuropathological and MRI study found that both lacunes and WMH identified by imaging showed good correlation with CVD (Jagust *et al.*, 2008).

Nonconventional MRI techniques, such as DWI (O'Sullivan *et al.*, 2004), MRS (Ross *et al.*, 2006), and MTI (Spilt *et al.*, 2006), may increase both the sensitivity and specificity of detecting histopathological changes that are not seen by conventional MRI sequences.

IV. Conclusions

Neuroimaging plays an important role in the assessment of cognitive impairment in vascular disease. Conventional MRI findings have been extensively investigated and correlated with cognitive function in vascular patients.

Despite recent advances, many unsolved problems and questions remain to be answered. In fact, conventional MRI findings of WMH are unspecific and might not differentiate between various grades of histopathological damage. Higher field strength MRI (3 T and more) and newer nonconventional techniques can provide important contributions toward better understanding of the pathophysiology of the disease and the relation between vascular lesions and cognitive function. Functional imaging provides important knowledge about how the ischemic lesions affect brain activity and may help clarify the brain's plasticity- and receptor-dependent compensatory mechanisms in patients with a variety of neurological disorders, including vascular diseases. Nonconventional imaging goes beyond MRI-visible pathology, providing more information about subtle brain damage in VCI. Its results suggest that better understanding of the pathophysiological aspects of the disease may be obtained in the future. These techniques are unique tools for *in vivo* assessment of VCI and are useful in establishing diagnosis, monitoring disease activity, measuring therapeutic effect, and explaining the development of disability in the short and long term. They should be extensively taken into consideration for developing a new and more unified classification of VCI. Further studies are warranted to clarify the correlation between vascular brain damage and cognitive impairment.

Acknowledgments

Dr Di Perri was supported by the Dr. Larry D. Jacobs Fellowship of the Buffalo Neuroimaging Analysis Center. Dr Dalaker was supported by the Dr. Larry D. Jacobs Fellowship of the Buffalo Neuroimaging Analysis Center and a grant from the Research Council of Norway (grant# 186966). Dr Beyer is supported by funds from the Norwegian Society of Radiology. The authors thank Eve Salczynski for technical assistance in the preparation of this manuscript.

References

Babikian, V., and Ropper, A. H. (1987). Binswanger's disease: A review. *Stroke* **18**(1), 2–12.

Barba, R., Martinez-Espinosa, S., Rodriguez-Garcia, E., Pondal, M., Vivancos, J., and Del Ser, T. (2000). Poststroke dementia: Clinical features and risk factors. *Stroke* **31**(7), 1494–1501.

Bombois, S., Debette, S., Delbeuck, X., Bruandet, A., Lepoittevin, S., Delmaire, C., Leys, D., and Pasquier, F. (2007). Prevalence of subcortical vascular lesions and association with executive function in mild cognitive impairment subtypes. *Stroke* **38**(9), 2595–2597.

Bombois, S., Debette, S., Bruandet, A., Delbeuck, X., Delmaire, C., Leys, D., and Pasquier, F. (2008). Vascular subcortical hyperintensities predict conversion to vascular and mixed dementia in MCI patients. *Stroke* **39,** 2046–2051.

Brown, W. R., Moody, D. M., Thore, C. R., Challa, V. R., and Anstrom, J. A. (2007). Vascular dementia in leukoaraiosis may be a consequence of capillary loss not only in the lesions, but in normal-appearing white matter and cortex as well. *J. Neurol. Sci.* **257**(1-2), 62–66.

Busto, R., Dietrich, W. D., Globus, M. Y., Valdes, I., Scheinberg, P., and Ginsberg, M. D. (1987). Small differences in intraischemic brain temperature critically determine the extent of ischemic neuronal injury. *J. Cereb. Blood Flow Metab.* **7**(6), 729–738.

Carey, C. L., Kramer, J. H., Josephson, S. A., Mungas, D., Reed, B. R., Schuff, N., Weiner, M. W., and Chui, H. C. (2008). Subcortical lacunes are associated with executive dysfunction in cognitively normal elderly. *Stroke* **39**(2), 397–402.

Chui, H. C., Victoroff, J. I., Margolin, D., Jagust, W., Shankle, R., and Katzman, R. (1992). Criteria for the diagnosis of ischemic vascular dementia proposed by the State of California Alzheimer's Disease Diagnostic and Treatment Centers. *Neurology* **42**(3 Pt 1), 473–480.

Clarke, S., Assal, G., Bogousslavsky, J., Regli, F., Townsend, D. W., Leenders, K. L., and Blecic, S. (1994). Pure amnesia after unilateral left polar thalamic infarct: Topographic and sequential neuropsychological and metabolic (PET) correlations. *J. Neurol. Neurosurg. Psychiatry* **57**(1), 27–34.

Cohen, R. A., Paul, R. H., Ott, B. R., Moser, D. J., Zawacki, T. M., Stone, W., and Gordon, N. (2002). The relationship of subcortical MRI hyperintensities and brain volume to cognitive function in vascular dementia. *J. Int. Neuropsychol. Soc.* **8**(6), 743–752.

Cordonnier, C., van der Flier, W. M., Sluimer, J. D., Leys, D., Barkhof, F., and Scheltens, P. (2006). Prevalence and severity of microbleeds in a memory clinic setting. *Neurology* **66**(9), 1356–1360.

Crain, M. R., Yuh, W. T., Greene, G. M., Loes, D. J., Ryals, T. J., Sato, Y., and Hart, M. N. (1991). Cerebral ischemia: Evaluation with contrast-enhanced MR imaging. *AJNR Am. J. Neuroradiol.* **12**(4), 631–639.

Cummings, J. L. (1994). Vascular subcortical dementias: Clinical aspects. *Dementia* **5**(3-4), 177–180.

de la Torre, J. C. (2004). Alzheimer's disease is a vasocognopathy: A new term to describe its nature. *Neurol. Res.* **26**(5), 517–524.

de Leeuw, F. E., de Groot, J. C., Achten, E., Oudkerk, M., Ramos, L. M., Heijboer, R., Hofman, A., Jolles, J., van Gijn, J., and Breteler, M. M. (2001). Prevalence of cerebral white matter lesions in elderly people: A population based magnetic resonance imaging study. The Rotterdam Scan Study. *J. Neurol. Neurosurg. Psychiatry* **70**(1), 9–14.

Debette, S., Bombois, S., Bruandet, A., Delbeuck, X., Lepoittevin, S., Delmaire, C., Leys, D., and Pasquier, F. (2007). Subcortical hyperintensities are associated with cognitive decline in patients with mild cognitive impairment. *Stroke* **38**(11), 2924–2930.

Decarli, C. (2004). Vascular factors in dementia: An overview. *J. Neurol. Sci.* **226**(1-2), 19–23.

Dichgans, M., Holtmannspotter, M., Herzog, J., Peters, N., Bergmann, M., and Yousry, T. A. (2002). Cerebral microbleeds in CADASIL: A gradient-echo magnetic resonance imaging and autopsy study. *Stroke* **33**(1), 67–71.

Enzinger, C., Fazekas, F., Matthews, P. M., Ropele, S., Schmidt, H., Smith, S., and Schmidt, R. (2005). Risk factors for progression of brain atrophy in aging: Six-year follow-up of normal subjects. *Neurology* **64**(10), 1704–1711.

Enzinger, C., Smith, S., Fazekas, F., Drevin, G., Ropele, S., Nichols, T., Behrens, T., Schmidt, R., and Matthews, P. M. (2006). Lesion probability maps of white matter hyperintensities in elderly individuals: Results of the Austrian stroke prevention study. *J. Neurol.* **253**(8), 1064–1070.

Erkinjuntti, T. (2007). Vascular cognitive deterioration and stroke. *Cerebrovasc. Dis.* **24**(Suppl. 1), 189–194.

Erkinjuntti, T., Kurz, A., Gauthier, S., Bullock, R., Lilienfeld, S., and Damaraju, C. V. (2002). Efficacy of galantamine in probable vascular dementia and Alzheimer's disease combined with cerebrovascular disease: A randomised trial. *Lancet* **359**(9314), 1283–1290.

Fazekas, F. (1989). Magnetic resonance signal abnormalities in asymptomatic individuals: Their incidence and functional correlates. *Eur. Neurol.* **29**(3), 164–168.

Fazekas, F., Kleinert, R., Offenbacher, H., Schmidt, R., Kleinert, G., Payer, F., Radner, H., and Lechner, H. (1993). Pathologic correlates of incidental MRI white matter signal hyperintensities. *Neurology* **43**(9), 1683–1689.

Fazekas, F., Kleinert, R., Roob, G., Kleinert, G., Kapeller, P., Schmidt, R., and Hartung, H. P. (1999). Histopathologic analysis of foci of signal loss on gradient-echo T2*-weighted MR images in patients with spontaneous intracerebral hemorrhage: Evidence of microangiopathy-related microbleeds. *AJNR Am. J. Neuroradiol.* **20**(4), 637–642.

Federico, A., Bianchi, S., and Dotti, M. T. (2005). The spectrum of mutations for CADASIL diagnosis. *Neurol. Sci.* **26**(2), 117–124.

Fein, G., Di Sclafani, V., Tanabe, J., Cardenas, V., Weiner, M. W., Jagust, W. J., Reed, B. R., Norman, D., Schuff, N., Kusdra, L., Greenfield, T., and Chui, H. (2000). Hippocampal and cortical atrophy predict dementia in subcortical ischemic vascular disease. *Neurology* **55**(11), 1626–1635.

Fernando, M. S., and Ince, P. G. (2004). Vascular pathologies and cognition in a population-based cohort of elderly people. *J. Neurol. Sci.* **226**(1-2), 13–17.

Fernando, M. S., Simpson, J. E., Matthews, F., Brayne, C., Lewis, C. E., Barber, R., Kalaria, R. N., Forster, G., Esteves, F., Wharton, S. B., Shaw, P. J., O'Brien, J. T., *et al.* (2006). White matter lesions in an unselected cohort of the elderly: Molecular pathology suggests origin from chronic hypoperfusion injury. *Stroke* **37**(6), 1391–1398.

Firbank, M. J., Burton, E. J., Barber, R., Stephens, S., Kenny, R. A., Ballard, C., Kalaria, R. N., and O'Brien, J. T. (2007). Medial temporal atrophy rather than white matter hyperintensities predict cognitive decline in stroke survivors. *Neurobiol. Aging* **28**(11), 1664–1669.

Gold, G., Kovari, E., Herrmann, F. R., Canuto, A., Hof, P. R., Michel, J. P., Bouras, C., and Giannakopoulos, P. (2005). Cognitive consequences of thalamic, basal ganglia, and deep white matter lacunes in brain aging and dementia. *Stroke* **36**(6), 1184–1188.

Gold, G., Giannakopoulos, P., Herrmann, F. R., Bouras, C., and Kovari, E. (2007a). Identification of Alzheimer and vascular lesion thresholds for mixed dementia. *Brain* **130**(Pt 11), 2830–2836.

Gold, G., Kovari, E., Hof, P. R., Bouras, C., and Giannakopoulos, P. (2007b). Sorting out the clinical consequences of ischemic lesions in brain aging: A clinicopathological approach. *J. Neurol. Sci.* **257**(1-2), 17–22.

Hachinski, V. C., Lassen, N. A., and Marshall, J. (1974). Multi-infarct dementia. A cause of mental deterioration in the elderly. *Lancet* **2**(7874), 207–210.

Hanyu, H., Tanaka, Y., Shimizu, S., Takasaki, M., and Abe, K. (2003). Cerebral microbleeds in Alzheimer's disease. *J. Neurol.* **250**(12), 1496–1497.

Horsfield, M. A., Barker, G. J., Barkhof, F., Miller, D. H., Thompson, A. J., and Filippi, M. (2003). Guidelines for using quantitative magnetization transfer magnetic resonance imaging for monitoring treatment of multiple sclerosis. *J. Magn. Reson. Imaging* **17**(4), 389–397.

Ishii, N., Nishihara, Y., and Imamura, T. (1986). Why do frontal lobe symptoms predominate in vascular dementia with lacunes? *Neurology* **36**(3), 340–345.

Jagust, W. J., Zheng, L., Harvey, D. J., Mack, W. J., Vinters, H. V., Weiner, M. W., Ellis, W. G., Zarow, C., Mungas, D., Reed, B. R., Kramer, J. H., Schuff, N., *et al.* (2008). Neuropathological basis of magnetic resonance images in aging and dementia. *Ann. Neurol.* **63**(1), 72–80.

Jellinger, K. A. (2007). The enigma of vascular cognitive disorder and vascular dementia. *Acta Neuropathol.* **113**(4), 349–388.

Kapeller, P., Schmidt, R., and Fazekas, F. (2004). Qualitative MRI: Evidence of usual aging in the brain. *Top. Magn. Reson. Imaging* **15**(6), 343–347.

Kerrouche, N., Herholz, K., Mielke, R., Holthoff, V., and Baron, J. C. (2006). 18FDG PET in vascular dementia: Differentiation from Alzheimer's disease using voxel-based multivariate analysis. *J. Cereb. Blood Flow Metab.* **26**(9), 1213–1221.

Korf, E. S., White, L. R., Scheltens, P., and Launer, L. J. (2006). Brain aging in very old men with type 2 diabetes: The Honolulu-Asia Aging Study. *Diabetes Care* **29**(10), 2268–2274.

Kovari, E., Gold, G., Herrmann, F. R., Canuto, A., Hof, P. R., Michel, J. P., Bouras, C., and Giannakopoulos, P. (2004). Cortical microinfarcts and demyelination significantly affect cognition in brain aging. *Stroke* **35**(2), 410–414.

Kushner, D. (1998). Mild traumatic brain injury: Toward understanding manifestations and treatment. *Arch Intern. Med.* **158**(15), 1617–1624.

Lesnik Oberstein, S. A., van Duinen, S. G., van den Boom, R., Maat-Schieman, M. L., van Buchem, M. A., van Houwelingen, H. C., Hegeman-Kleinn, I. M., Ferrari, M. D., Breuning, M. H., and Haan, J. (2003). Evaluation of diagnostic NOTCH3 immunostaining in CADASIL. *Acta Neuropathol.* **106**(2), 107–111.

Leys, D., Henon, H., Mackowiak-Cordoliani, M. A., and Pasquier, F. (2005). Poststroke dementia. *Lancet Neurol.* **4**(11), 752–759.

Liao, D., Cooper, L., Cai, J., Toole, J., Bryan, N., Burke, G., Shahar, E., Nieto, J., Mosley, T., and Heiss, G. (1997). The prevalence and severity of white matter lesions, their relationship with age, ethnicity, gender, and cardiovascular disease risk factors: The ARIC Study. *Neuroepidemiology* **16**(3), 149–162.

Libon, D. J., Price, C. C., Giovannetti, T., Swenson, R., Bettcher, B. M., Heilman, K. M., and Pennisi, A. (2008). Linking MRI hyperintensities with patterns of neuropsychological impairment: Evidence for a threshold effect. *Stroke* **39**(3), 806–813.

Liem, M. K., van der Grond, J., Haan, J., van den Boom, R., Ferrari, M. D., Knaap, Y. M., Breuning, M. H., van Buchem, M. A., Middelkoop, H. A., and Lesnik Oberstein, S. A. (2007). Lacunar infarcts are the main correlate with cognitive dysfunction in CADASIL. *Stroke* **38**(3), 923–928.

Linden, T., Skoog, I., Fagerberg, B., Steen, B., and Blomstrand, C. (2004). Cognitive impairment and dementia 20 months after stroke. *Neuroepidemiology* **23**(1-2), 45–52.

Lojkowska, W., Ryglewicz, D., Jedrzejczak, T., Sienkiewicz-Jarosz, H., Minc, S., Jakubowska, T., and Kozlowicz-Gudzinska, I. (2002). SPECT as a diagnostic test in the investigation of dementia. *J. Neurol. Sci.* **203-204,** 215–219.

Lovblad, K. O., Baird, A. E., Schlaug, G., Benfield, A., Siewert, B., Voetsch, B., Connor, A., Burzynski, C., Edelman, R. R., and Warach, S. (1997). Ischemic lesion volumes in acute stroke by diffusion-weighted magnetic resonance imaging correlate with clinical outcome. *Ann. Neurol.* **42**(2), 164–170.

Maxeiner, H., and Wolff, M. (2007). Pure subdural hematomas: A postmortem analysis of their form and bleeding points. *Neurosurgery* **61**(1 Suppl.), 267-272; discussion 272–273.

Meyer, J. S., Rogers, R. L., McClintic, K., Mortel, K. F., and Lotfi, J. (1989). Randomized clinical trial of daily aspirin therapy in multi-infarct dementia. A pilot study. *J. Am. Geriatr. Soc.* **37**(6), 549–555.

Mielke, R., Pietrzyk, U., Jacobs, A., Fink, G. R., Ichimiya, A., Kessler, J., Herholz, K., and Heiss, W. D. (1994). HMPAO SPET and FDG PET in Alzheimer's disease and vascular dementia: Comparison of perfusion and metabolic pattern. *Eur. J. Nucl. Med.* **21**(10), 1052–1060.

Murray, A. D., Staff, R. T., Shenkin, S. D., Deary, I. J., Starr, J. M., and Whalley, L. J. (2005). Brain white matter hyperintensities: Relative importance of vascular risk factors in nondemented elderly people. *Radiology* **237**(1), 251–257.

Nagata, K., Maruya, H., Yuya, H., Terashi, H., Mito, Y., Kato, H., Sato, M., Satoh, Y., Watahiki, Y., Hirata, Y., Yokoyama, E., and Hatazawa, J. (2000). Can PET data differentiate Alzheimer's disease from vascular dementia? *Ann. NY Acad. Sci.* **903,** 252–261.

Nagata, K., Saito, H., Ueno, T., Sato, M., Nakase, T., Maeda, T., Satoh, Y., Komatsu, H., Suzuki, M., and Kondoh, Y. (2007). Clinical diagnosis of vascular dementia. *J. Neurol. Sci.* **257**(1-2), 44–48.

Narayana, P. A. (2005). Magnetic resonance spectroscopy in the monitoring of multiple sclerosis. *J. Neuroimaging* **15**(4 Suppl.), 46S–57S.

Narayanan, S., Francis, S. J., Sled, J. G., Santos, A. C., Antel, S., Levesque, I., Brass, S., Lapierre, Y., Sappey-Marinier, D., Pike, G. B., and Arnold, D. L. (2006). Axonal injury in the cerebral normal-appearing white matter of patients with multiple sclerosis is related to concurrent demyelination in lesions but not to concurrent demyelination in normal-appearing white matter. *Neuroimage* **29**(2), 637–642.

Nicoll, J. A., Yamada, M., Frackowiak, J., Mazur-Kolecka, B., and Weller, R. O. (2004). Cerebral amyloid angiopathy plays a direct role in the pathogenesis of Alzheimer's disease. Pro-CAA position statement. *Neurobiol. Aging* **25**(5), 589–597; discussion 603-4.

O'Brien, J. T. (2006). Vascular cognitive impairment. *Am. J. Geriatr. Psychiatry* **14**(9), 724–733.

O'Brien, J. T., Erkinjuntti, T., Reisberg, B., Roman, G., Sawada, T., Pantoni, L., Bowler, J. V., Ballard, C., DeCarli, C., Gorelick, P. B., Rockwood, K., Burns, A., *et al.* (2003). Vascular cognitive impairment. *Lancet Neurol.* **2**(2), 89–98.

O'Sullivan, M., Morris, R. G., Huckstep, B., Jones, D. K., Williams, S. C., and Markus, H. S. (2004). Diffusion tensor MRI correlates with executive dysfunction in patients with ischaemic leukoaraiosis. *J. Neurol. Neurosurg. Psychiatry* **75**(3), 441–447.

O'Sullivan, M., Ngo, E., Viswanathan, A., Jouvent, E., Gschwendtner, A., Saemann, P. G., Duering, M., Pachai, C., Bousser, M. G., Chabriat, H., and Dichgans, M. (2007). Hippocampal volume is an independent predictor of cognitive performance in CADASIL. *Neurobiol. Aging.* In press.

Peters, N., Holtmannspotter, M., Opherk, C., Gschwendtner, A., Herzog, J., Samann, P., and Dichgans, M. (2006). Brain volume changes in CADASIL: A serial MRI study in pure subcortical ischemic vascular disease. *Neurology* **66**(10), 1517–1522.

Petersen, R. C., Smith, G. E., Waring, S. C., Ivnik, R. J., Tangalos, E. G., and Kokmen, E. (1999). Mild cognitive impairment: Clinical characterization and outcome. *Arch. Neurol.* **56**(3), 303–308.

Pohjasvaara, T., Mantyla, R., Salonen, O., Aronen, H. J., Ylikoski, R., Hietanen, M., Kaste, M., and Erkinjuntti, T. (2000). How complex interactions of ischemic brain infarcts, white matter lesions, and atrophy relate to poststroke dementia. *Arch. Neurol.* **57**(9), 1295–1300.

Pratt, R. D., and Perdomo, C. A. (2002). Donepezil-treated patients with probable vascular dementia demonstrate cognitive benefits. *Ann. NY Acad. Sci.* **977,** 513–522.

Reed, B. R., Eberling, J. L., Mungas, D., Weiner, M., Kramer, J. H., and Jagust, W. J. (2004). Effects of white matter lesions and lacunes on cortical function. *Arch. Neurol.* **61**(10), 1545–1550.

Rockwood, K., Wentzel, C., Hachinski, V., Hogan, D. B., MacKnight, C., and McDowell, I. (2000). Prevalence and outcomes of vascular cognitive impairment. Vascular Cognitive Impairment Investigators of the Canadian Study of Health and Aging. *Neurology* **54**(2), 447–451.

Roman, G. C., Tatemichi, T. K., Erkinjuntti, T., Cummings, J. L., Masdeu, J. C., Garcia, J. H., Amaducci, L., Orgogozo, J. -M., Brun, A., Hofman, A., Moody, D. M., O'Brien, M. D., *et al.* (1993). Vascular dementia: Diagnostic criteria for research studies. Report of the NINDS-AIREN International Workshop. *Neurology* **43**(2), 250–260.

Roman, G. C., Sachdev, P., Royall, D. R., Bullock, R. A., Orgogozo, J. M., Lopez-Pousa, S., Arizaga, R., and Wallin, A. (2004). Vascular cognitive disorder: A new diagnostic category updating vascular cognitive impairment and vascular dementia. *J. Neurol. Sci.* **226**(1-2), 81–87.

Ross, A. J., Sachdev, P. S., Wen, W., Brodaty, H., Joscelyne, A., and Lorentz, L. M. (2006). Prediction of cognitive decline after stroke using proton magnetic resonance spectroscopy. *J. Neurol. Sci.* **251**(1-2), 62–69.

Rothwell, P. M., Coull, A. J., Giles, M. F., Howard, S. C., Silver, L. E., Bull, L. M., Gutnikov, S. A., Edwards, P., Mant, D., Sackley, C. M., Farmer, A., Sandercock, P. A., *et al.* (2004). Change in stroke incidence, mortality, case-fatality, severity, and risk factors in Oxfordshire, UK from 1981 to 2004 (Oxford Vascular Study). *Lancet* **363**(9425), 1925–1933.

Rovaris, M., Gass, A., Bammer, R., Hickman, S. J., Ciccarelli, O., Miller, D. H., and Filippi, M. (2005). Diffusion MRI in multiple sclerosis. *Neurology* **65**(10), 1526–1532.

Sachdev, P. S., Wen, W., Christensen, H., and Jorm, A. F. (2005). White matter hyperintensities are related to physical disability and poor motor function. *J. Neurol. Neurosurg. Psychiatry* **76**(3), 362–367.

Sachdev, P. S., Brodaty, H., Valenzuela, M. J., Lorentz, L., Looi, J. C., Berman, K., Ross, A., Wen, W., and Zagami, A. S. (2006). Clinical determinants of dementia and mild cognitive impairment following ischaemic stroke: The Sydney Stroke Study. *Dement. Geriatr. Cogn. Disord* **21**(5-6), 275–283.

Sachdev, P. S., Parslow, R., Wen, W., Anstey, K. J., and Easteal, S. (2007). Sex differences in the causes and consequences of white matter hyperintensities. *Neurobiol. Aging.* In Press.

Saha, G. (2006). "Physics and Radiobiology of Nuclear Medicine," 3rd edition. Springer, New York.

Schmidt, R., Petrovic, K., Ropele, S., Enzinger, C., and Fazekas, F. (2007). Progression of leukoaraiosis and cognition. *Stroke* **38**(9), 2619–2625.

Schmidtke, K., and Hull, M. (2005). Cerebral small vessel disease: How does it progress? *J. Neurol. Sci.* **229-230,** 13–20.

Silverman, D. H. (2004). Brain 18F-FDG PET in the diagnosis of neurodegenerative dementias: Comparison with perfusion SPECT and with clinical evaluations lacking nuclear imaging. *J. Nucl. Med.* **45**(4), 594–607.

Skehan, S. J., Hutchinson, M., and MacErlaine, D. P. (1995). Cerebral autosomal dominant arteriopathy with subcortical infarcts and leukoencephalopathy: MR findings. *AJNR Am. J. Neuroradiol.* **16**(10), 2115–2119.

Skoog, I., Andreasson, L. A., Landahl, S., and Lernfelt, B. (1998). A population-based study on blood pressure and brain atrophy in 85-year-olds. *Hypertension* **32**(3), 404–409.

Sobesky, J., Zaro Weber, O., Lehnhardt, F. G., Hesselmann, V., Neveling, M., Jacobs, A., and Heiss, W. D. (2005). Does the mismatch match the penumbra? Magnetic resonance imaging and positron emission tomography in early ischemic stroke. *Stroke* **36**(5), 980–985.

Spilt, A., Goekoop, R., Westendorp, R. G., Blauw, G. J., de Craen, A. J., and van Buchem, M. A. (2006). Not all age-related white matter hyperintensities are the same: A magnetization transfer imaging study. *AJNR Am. J. Neuroradiol.* **27**(9), 1964–1968.

Starkstein, S. E., Jorge, R., and Capizzano, A. A. (2005). Uncommon causes of cerebrovascular dementia. *Int. Psychogeriatr.* **17**(Suppl. 1), S51–64.

Sultzer, D. L., Mahler, M. E., Cummings, J. L., Van Gorp, W. G., Hinkin, C. H., and Brown, C. (1995). Cortical abnormalities associated with subcortical lesions in vascular dementia. Clinical and position emission tomographic findings. *Arch. Neurol.* **52**(8), 773–780.

Sweet, L. H., Paul, R. H., Cohen, R. A., Moser, D., Ott, B. R., Gordon, N., Browndyke, J. N., Shah, P., and Garrett, K. D. (2003). Neuroimaging correlates of dementia rating scale performance at baseline and 12-month follow-up among patients with vascular dementia. *J. Geriatr. Psychiatry Neurol.* **16**(4), 240–244.

Takahashi, W., Takagi, S., Ide, M., Shohtsu, A., and Shinohara, Y. (2000). Reduced cerebral glucose metabolism in subjects with incidental hyperintensities on magnetic resonance imaging. *J. Neurol. Sci.* **176**(1), 21–27.

Tatemichi, T. K., Foulkes, M. A., Mohr, J. P., Hewitt, J. R., Hier, D. B., Price, T. R., and Wolf, P. A. (1990). Dementia in stroke survivors in the Stroke Data Bank cohort. Prevalence, incidence, risk factors, and computed tomographic findings. *Stroke* **21**(6), 858–866.

Tatemichi, T. K., Desmond, D. W., Mayeux, R., Paik, M., Stern, Y., Sano, M., Remien, R. H., Williams, J. B., Mohr, J. P., Hauser, W. A., *et al.* (1992). Dementia after stroke: Baseline frequency, risks, and clinical features in a hospitalized cohort. *Neurology* **42**(6), 1185–1193.

Tian, M., Zhang, H., Oriuchi, N., Higuchi, T., and Endo, K. (2004). Comparison of 11C-choline PET and FDG PET for the differential diagnosis of malignant tumors. *Eur. J. Nucl. Med. Mol. Imaging* **31**(8), 1064–1072.

van den Heuvel, D. M., Admiraal-Behloul, F., ten Dam, V. H., Olofsen, H., Bollen, E. L., Murray, H. M., Blauw, G. J., Westendorp, R. G., de Craen, A. J., and van Buchem, M. A. (2004). Different progression rates for deep white matter hyperintensities in elderly men and women. *Neurology* **63**(9), 1699–1701.

van der Flier, W. M., van Straaten, E. C., Barkhof, F., Verdelho, A., Madureira, S., Pantoni, L., Inzitari, D., Erkinjuntti, T., Crisby, M., Waldemar, G., Schmidt, R., Fazekas, F., *et al.* (2005). Small vessel disease and general cognitive function in nondisabled elderly: The LADIS study. *Stroke* **36**(10), 2116–2120.

Vataja, R., Pohjasvaara, T., Mantyla, R., Ylikoski, R., Leppavuori, A., Leskela, M., Kalska, H., Hietanen, M., Aronen, H. J., Salonen, O., Kaste, M., and Erkinjuntti, T. (2003). MRI correlates of executive dysfunction in patients with ischaemic stroke. *Eur. J. Neurol.* **10**(6), 625–631.

Verdelho, A., Madureira, S., Ferro, J. M., Basile, A. M., Chabriat, H., Erkinjuntti, T., Fazekas, F., Hennerici, M., O'Brien, J., Pantoni, L., Salvadori, E., Scheltens, P., *et al.* (2007). Differential impact of cerebral white matter changes, diabetes, hypertension and stroke on cognitive performance among non-disabled elderly. The LADIS study. *J. Neurol. Neurosurg. Psychiatry* **78**(12), 1325–1330.

Viswanathan, A., Gschwendtner, A., Guichard, J. P., Buffon, F., Cumurciuc, R., O'Sullivan, M., Holtmannspotter, M., Pachai, C., Bousser, M. G., Dichgans, M., and Chabriat, H. (2007). Lacunar lesions are independently associated with disability and cognitive impairment in CADASIL. *Neurology* **69**(2), 172–179.

Vrenken, H., Barkhof, F., Uitdehaag, B. M., Castelijns, J. A., Polman, C. H., and Pouwels, P. J. (2005). MR spectroscopic evidence for glial increase but not for neuro-axonal damage in MS normal-appearing white matter. *Magn. Reson. Med.* **53**(2), 256–266.

Wallin, A., Milos, V., Sjogren, M., Pantoni, L., and Erkinjuntti, T. (2003). Classification and subtypes of vascular dementia. *Int. Psychogeriatr.* **15**(Suppl. 1), 27–37.

Wen, W., and Sachdev, P. (2004). The topography of white matter hyperintensities on brain MRI in healthy 60- to 64-year-old individuals. *Neuroimage* **22**(1), 144–154.

Wen, W., Sachdev, P. S., Chen, X., and Anstey, K. (2006). Gray matter reduction is correlated with white matter hyperintensity volume: A voxel-based morphometric study in a large epidemiological sample. *Neuroimage* **29**(4), 1031–1039.

Wentzel, C., Rockwood, K., MacKnight, C., Hachinski, V., Hogan, D. B., Feldman, H., Ostbye, T., Wolfson, C., Gauthier, S., Verreault, R., and McDowell, I. (2001). Progression of impairment in patients with vascular cognitive impairment without dementia. *Neurology* **57**(4), 714–716.

Werring, D. J., Frazer, D. W., Coward, L. J., Losseff, N. A., Watt, H., Cipolotti, L., Brown, M. M., and Jager, H. R. (2004). Cognitive dysfunction in patients with cerebral microbleeds on T2*-weighted gradient-echo MRI. *Brain* **127**(Pt 10), 2265–2275.

Whitwell, J. L. (2008). Longitudinal imaging: Change and causality. *Curr. Opin. Neurol.* **21**(4), 410–416.

Wolpert, S. M., Bruckmann, H., Greenlee, R., Wechsler, L., Pessin, M. S., and del Zoppo, G. J. (1993). Neuroradiologic evaluation of patients with acute stroke treated with recombinant tissue plasminogen activator. The rt-PA Acute Stroke Study Group. *AJNR Am. J. Neuroradiol.* **14**(1), 3–13.

Won Seo, S., Hwa Lee, B., Kim, E. J., Chin, J., Sun Cho, Y., Yoon, U., and Na, D. L. (2007). Clinical significance of microbleeds in subcortical vascular dementia. *Stroke* **38**(6), 1949–1951.

Wright, C. B., Festa, J. R., Paik, M. C., Schmiedigen, A., Brown, T. R., Yoshita, M., DeCarli, C., Sacco, R., and Stern, Y. (2008). White matter hyperintensities and subclinical infarction: Associations with psychomotor speed and cognitive flexibility. *Stroke* **39**(3), 800–805.

Yang, D. W., Kim, B. S., Park, J. K., Kim, S. Y., Kim, E. N., and Sohn, H. S. (2002). Analysis of cerebral blood flow of subcortical vascular dementia with single photon emission computed tomography: Adaptation of statistical parametric mapping. *J. Neurol. Sci.* **203-204,** 199–205.

Yoshita, M., Fletcher, E., Harvey, D., Ortega, M., Martinez, O., Mungas, D. M., Reed, B. R., and DeCarli, C. S. (2006). Extent and distribution of white matter hyperintensities in normal aging, MCI, and AD. *Neurology* **67**(12), 2192–2198.

Zhu, L., Viitanen, M., Guo, Z., Winblad, B., and Fratiglioni, L. (1998). Blood pressure reduction, cardiovascular diseases, and cognitive decline in the mini-mental state examination in a community population of normal very old people: A three-year follow-up. *J. Clin. Epidemiol.* **51**(5), 385–391.

Zimny, A., Leszek, J., Kiejna, A., and Sasiadek, M. (2007). Analysis of correlation between the degree of cognitive impairment and the results of perfusion CT in patients with dementia. *Med. Sci. Monit.* **13**(Suppl. 1), 23–30.

Zivadinov, R., and Bakshi, R. (2004). Role of MRI in multiple sclerosis I: Inflammation and lesions. *Front Biosci.* **9,** 665–683.

Zivadinov, R., and Cox, J. L. (2007). Neuroimaging in multiple sclerosis. *Int. Rev. Neurobiol.* **79,** 449–474.

CONTRIBUTIONS OF NEUROPSYCHOLOGY AND NEUROIMAGING TO UNDERSTANDING CLINICAL SUBTYPES OF MILD COGNITIVE IMPAIRMENT

Amy J. Jak,[*,†] Katherine J. Bangen,[*] Christina E. Wierenga,[*,†] Lisa Delano-Wood,[*,†] Jody Corey-Bloom,[†,‡] and Mark W. Bondi[*,†]

[*]Department of Psychiatry, School of Medicine, University of California, San Diego 92093, USA

[†]Veterans Affairs San Diego Healthcare System, San Diego 92161, USA

[‡]Department of Neurosciences, School of Medicine, University of California, San Diego 92093, USA

The original conceptualization of mild cognitive impairment (MCI) was primarily as an amnestic disorder representing an intermediate stage between normal aging and Alzheimer's dementia (AD). More recently, broader conceptualizations of MCI have emerged that also encompass cognitive domains other than memory. These characterizations delineate clinical subtypes that commonly include amnestic and non-amnestic forms, and that involve single and multiple cognitive domains. With the advent of these broader classifications, more specific information is emerging regarding the neuropsychological presentation of individuals with MCI, risk for dementia associated with different subtypes of MCI, and neuropathologic substrates connected to the clinical subtypes. This review provides an overview of this burgeoning literature specific to clinical subtypes of MCI. Focus is primarily on neuropsychological and structural neuroimaging

INTERNATIONAL REVIEW OF NEUROBIOLOGY, VOL. 84
DOI: 10.1016/S0074-7742(09)00405-X

findings specific to clinical subtypes of MCI as well as the issue of daily functioning. Although investigations of non-amnestic subtypes using advanced neuroimaging techniques and clinical trials are quite limited, we briefly review these topics in MCI because these data provide a framework for future investigations specifically examining additional clinical subtypes of MCI. Finally, the review comments on select methodological issues involved in studying this heterogeneous population, and future directions to continue to improve our understanding of MCI and its clinical subtypes are offered.

I. Introduction

Mild cognitive impairment (MCI) is a clinical construct that describes individuals with mildly impaired performance on objective neuropsychological tests but relatively intact global cognition and daily functioning (Petersen *et al.*, 1999, 2001). MCI has been validated as qualitatively different from both normal aging and dementia (Petersen, 2004; Smith and Ivnik, 2003) and is a risk factor for the development of dementia. Because of its potential importance for early identification and intervention in those at risk for the development of dementia, the concept of MCI has received considerable research attention. However, the definition has evolved considerably over time. As originally proposed by Petersen and colleagues, MCI was characterized primarily as an amnestic disorder that represented an intermediate stage between normal aging and Alzheimer's dementia (AD) (Petersen *et al.*, 1999). More recently, broader conceptualizations of MCI have emerged that also encompass cognitive domains other than memory (Petersen and Morris, 2005; Petersen *et al.*, 2001). These characterizations delineate clinical subtypes that commonly include amnestic and non-amnestic forms, and that involve single and multiple cognitive domains (Manly *et al.*, 2005; Petersen and Morris, 2005; Petersen *et al.*, 2001; Tabert *et al.*, 2006) against the backdrop of intact daily functioning. With the advent of these broader classifications schemes, more specific information is emerging regarding the neuropsychological presentation of individuals with MCI, risk for dementia associated with different subtypes of MCI, daily functioning, and neuropathologic substrates connected to the clinical subtypes. The aim of this review is to provide an overview of neuropsychological, neuroimaging, functional, and treatment findings specific to clinical subtypes of MCI. In addition, methodological issues involved in studying this heterogeneous population and future directions to continue to improve our understanding of MCI and its clinical subtypes will also be highlighted and discussed.

II. Neuropsychological Presentation

It is certainly of great interest to determine factors placing individuals at highest risk for development of dementia so as to target them for early intervention. To this end, a better understanding of who is at risk for developing MCI may be an important first step. Presently, there are limited data about risk factors that correspond to conversion from cognitively normal to specific clinical subtypes of MCI; although, the existing evidence suggests that advancing age and lower education levels do place individuals at higher risk for MCI (Kryscio *et al.*, 2006), particularly for non-amnestic subtypes (Bickel *et al.*, 2006; Fischer *et al.*, 2007). For example, individuals with less than 9 years of education have an increased likelihood of isolated language and visuospatial deficits, as well as multiple domain amnestic MCI, whereas higher education was associated with increased chance of having isolated memory and executive impairments (Manly *et al.*, 2005). The presence of the apolipoprotein epsilon 4 allele (APOE $\varepsilon 4$) seems to more strongly influence transitions from normal cognition to amnestic MCI and influence conversion to multi-domain MCI to a lesser degree (Kryscio *et al.*, 2006).

More research attention has been paid to the neuropsychological presentation of MCI and use of neuropsychological testing to delineate distinct clinical subtypes (e.g., amnestic vs non-amnestic and single domain vs multiple domain). Even when stratified into clinical subtypes, MCI is still a heterogeneous concept. Complicating factors include widely differing neuropsychological tests and diagnostic criteria used across studies in arriving at the MCI classifications as well as inconsistency in how clinical subtypes are assigned. While Petersen advocated for four subtypes, including single and multiple domain amnestic MCI and single and multiple domain non-amnestic MCI, non-amnestic subtypes continue to receive less attention in the literature and multiple domain classifications often do not separate amnestic from non-amnestic presentations. Despite these challenges, some converging evidence about the presentation of distinct MCI subtypes is emerging.

Generally, multi-domain presentations seem to be more common than purely amnestic presentations (Alexopoulos *et al.*, 2006; Alladi *et al.*, 2006; Lopez *et al.*, 2003; Manly *et al.*, 2005; Rasquin *et al.*, 2005); although, some studies have identified single domains that are more common than multi-domains, and the non-amnestic type is as frequent as the amnestic (Busse *et al.*, 2006; Palmer *et al.*, 2008). Multi-domain MCI (mMCI) is the most common subtype in both stroke and memory clinic samples (Rasquin *et al.*, 2005). The prevalence of single domain amnestic MCI (aMCI) ranges from 0.5 to 8% (Bickel *et al.*, 2006; Das *et al.*, 2007; Di Carlo *et al.*, 2007; Jungwirth *et al.*, 2005; Lopez *et al.*, 2003), mMCI ranges from 0.5 to 16% (Bickel *et al.*, 2006; Busse *et al.*, 2003; Das *et al.*, 2007;

Di Carlo *et al.*, 2007; Jungwirth *et al.*, 2005; Lopez *et al.*, 2003), and single domain non-amnestic MCI ranges from approximately 3 to 15% (Bickel *et al.*, 2006; Busse *et al.*, 2003; Di Carlo *et al.*, 2007). There is only limited information on prevalence rates of multi-domain non-amnestic MCI, in which one group has identified less than a 5% prevalence rate (Bickel *et al.*, 2006). In addition to vastly different methodological approaches to defining MCI, the prevalence rates also differ due to varying sample origins, with hospital samples having generally higher prevalence rates across subtypes than community samples (Bickel *et al.*, 2006; Busse *et al.*, 2003). Presence of the APOE ε4 allele may also contribute to differing prevalence rates. Amnestic MCI presentations have a higher proportion of individuals with the apolipoprotein ε4 allele as compared to non-amnestic groups (Gabryelewicz *et al.*, 2007; Manly *et al.*, 2005; Whitwell *et al.*, 2007). Thus, if genetic risk is not determined, samples are likely to have differing proportions of individuals with the ε4 allele and, therefore, different prevalence rates.

Certainly the prevalence rates are impacted by the definition of MCI applied by each study. Basing amnestic subtype diagnoses on the presence of either verbal and/or visual memory deficits results in a larger proportion of individuals identified as amnestic MCI than just relying on verbal memory alone (Alladi *et al.*, 2006). Varying the cutoff score for defining impairment also alters diagnostic outcomes by up to 12%; use of a more stringent statistical cutoff for impairment (1.5–2.0 SD below normative expectations) increases positive predictive power compared to lower cut points (Busse *et al.*, 2003); although, a more liberal cutoff for impairment has been shown to have higher sensitivity and specificity for future development of dementia (Busse *et al.*, 2006). These sensitivity and specificity determinations are problematical; however, as they are based on a quite limited neuropsychological assessment.

Aside from the use in diagnosis and determination of objective cognitive deficits in MCI, additional neuropsychological findings may help differentiate MCI subtypes. Perhaps not surprisingly, Lopez and colleagues (2006) found that, compared to aMCI and normal cognition, mMCI was characterized by poorer language, psychomotor speed, fine motor control, and visuoconstructional functioning. What is of note is that, although the mMCI group had memory deficits, they were to a lesser degree than the deficits noted in the aMCI group (Lopez *et al.*, 2006). A substantial minority of the MCI cases did not have any memory impairment (Lopez *et al.*, 2006), further emphasizing that examining only amnestic subtypes fails to capture the full spectrum of possible cognitive declines associated with MCI.

Spatial navigation skills of those with multi-domain amnestic MCI tend to be more similar to spatial navigation skills of AD patients than to non-amnestic MCI subtype groups, with both the AD and multi-domain amnestic MCI groups impaired on virtually all portions of a spatial navigation task (Hort *et al.*, 2007). However, the multi-domain amnestic MCI group was generally more impaired than other groups across all neuropsychological tests, so it is not clear that these spatial navigation difficulties occurred in isolation from the other impaired functions. Visuospatial skills specific to facial emotional processing have also been found to be intact in those with single domain amnestic MCI but impaired in those with multi-domain amnestic MCI, particularly in facial affect discrimination (Teng *et al.*, 2007).

III. Stability of Diagnosis

Multiple studies indicate that not all individuals diagnosed with MCI will decline and progress to a dementia diagnosis. In fact, a proportion of individuals appear to "improve" over time such that, at follow up, those initially identified as MCI are later categorized as cognitively normal. Anywhere from 20 (Fischer *et al.*, 2007) to 40% (Bickel *et al.*, 2006) of those with MCI appear to revert to the normal range upon retesting. Single domain classifications appear particularly susceptible to this instability, with single domain non-amnestic subtype often exhibiting the least stability over time (Bickel *et al.*, 2006; Busse *et al.*, 2006; Fischer *et al.*, 2007; Jak *et al.*, 2007). For example, one study reported that 50% of those with single domain MCI were normal upon later retesting whereas only 12% of those with multi-domain MCI "recovered" (Bickel *et al.*, 2006). Additional sources of instability in the MCI diagnosis can manifest via individuals changing MCI subtypes over time. Approximately 6% of those with MCI at baseline changed subtypes at follow up (Fischer *et al.*, 2007; Jak *et al.*, 2007). In contrast, over a 3-year interval, Zanetti and colleagues (2006) identified a more anticipated trajectory of their MCI cohort; all MCI subtypes either converted to dementia (about a quarter) or retained their MCI status (Zanetti *et al.*, 2006).

IV. Conversion to Dementia

Perhaps the largest amount of information exists on likelihood of conversion to dementia from various MCI clinical subtypes. Some evidence suggests that those with multi-domain amnestic MCI appear to be at greatest risk for future dementia (Di Carlo *et al.*, 2007; Palmer *et al.*, 2008; Tabert *et al.*, 2006), whereas others indicate that amnestic MCI places one at highest risk for conversion to dementia (Ravaglia *et al.*, 2006; Yaffe *et al.*, 2006). Amnestic MCI subtypes do seem to impart significant risk for future development of AD while multi-domain presentations may be more common in those who eventually develop vascular dementia (Fischer *et al.*, 2007; Rasquin *et al.*, 2005; Yaffe *et al.*, 2006; Zanetti *et al.*, 2006). Yaffe and colleagues (2006) found that, of those who progressed to AD,

76% were initially diagnosed with aMCI, 11% initially presented with single domain non-amnestic MCI, and 13% were initially identified as mMCI (Yaffe *et al.*, 2006). Conversely, of those who progressed to vascular dementia, 50% were initially diagnosed with aMCI, 8% had single domain non-amnestic MCI, and 42% had mMCI (Yaffe *et al.*, 2006). Rozzini and colleagues (2007) reported that, in a group of amnestic MCI individuals, poor global cognitive performance at baseline and worsening executive functioning, but not worsening memory performance, were associated with conversion to AD over a 1-year follow-up period. Those with non-amnestic multiple domain subtype appear more likely to convert to a non-AD dementia (Busse *et al.*, 2006), with the single domain non-amnestic MCI at particular risk to progress to a frontal dementia syndrome (Yaffe *et al.*, 2006). There are reports, however, in which detailed information about MCI subtypes does not add significant benefit in determining who may be at greatest risk for conversion to dementia (Maioli *et al.*, 2007; Ravaglia *et al.*, 2006; Rountree *et al.*, 2007).

V. MCI and Health Variables

Understanding any additional health factors that may be more prevalent in distinct MCI subtypes is also noteworthy as a way to further delineate risk profiles. For example, cardiovascular risk factors, presence of the apolipoprotein ε4 allele, mood symptoms, and parkinsonian symptoms have all been investigated in MCI subtypes. Recent research has shown that multi-domain or non-amnestic MCI subtypes may be more likely to have cardiovascular risk factors than either those with single domain amnestic presentations or those without MCI (Di Carlo *et al.*, 2007; Zanetti *et al.*, 2006). Mariani and colleagues (2007) found that those with single domain non-amnestic MCI had a higher frequency of ischemic heart disease, transient ischemic attack (TIA) or stroke, a higher Hachinski ischemic score, and more white-matter lesions on MRI compared to aMCI. Further, multi-domain and single domain amnestic subjects exhibited similar clinical characteristics; although, the multi-domain amnestic subtype did have a greater history of TIA/stroke (Mariani *et al.*, 2007). Amnestic MCI groups showed a higher prevalence of diabetes than controls whereas participants with non-amnestic MCI were more likely to have hypertension than were controls (Verghese *et al.*, 2008).

In contrast, Debette *et al.* (2007) found that white matter changes may play a role in cognitive decline in MCI as a whole, but they do not appear to be specific to either amnestic or non-amnestic clinical subtypes. Because their amnestic and non-amnestic characterizations were multi-domain, the authors found the rate of cognitive decline and the presence of periventricular hyperintensities was more prominent in those with baseline executive dysfunction (Debette *et al.*, 2007).

Mood symptoms also appear to be more common in those with multi-domain MCI relative to those with single domain amnestic presentations (Gabryelewicz *et al.*, 2007; Zanetti *et al.*, 2006). Additionally, higher depression ratings are linked to those whose cognitive symptoms progressed over the follow-up period (but not to the point of dementia) and to those who converted from MCI to dementia as compared to the stable group (Gabryelewicz *et al.*, 2007). Hallucinations and sleep disorders are more common in the non-amnestic than amnestic subtype (Rozzini *et al.*, 2008). Others have found, however, that mood symptoms are increased in MCI cohorts as compared to cognitively normal samples, but with no significant differences in mood symptoms emerging between specific MCI subtypes (Palmer *et al.*, 2007; Rozzini *et al.*, 2008).

Other health factors seem to be associated with MCI in general but are not necessarily specific to amnestic or non-amnestic presentations. For example, lower serum folate levels (Maioli *et al.*, 2007; Ravaglia *et al.*, 2006), history of atrial fibrillation (Ravaglia *et al.*, 2006), and higher serum HDL levels (Maioli *et al.*, 2007) contribute to increased likelihood of conversion from MCI to dementia, regardless of MCI subtype. Odor identification skills of MCI participants also fall in between those of AD and healthy control groups, and MCI subtypes did not differ on smell identification performances (Westervelt *et al.*, 2008).

Most conceptualizations of MCI exclude Parkinson's disease, given that the historical conceptualization has been of MCI as a precursor to Alzheimer's disease or other non-Parkinsonian dementias. In addition, the motor symptoms associated with Parkinson's disease often produce demonstrable changes in a person's activities of daily living (ADL), which confound its utility in the classification of MCI. Recently, however, there have been efforts to characterize the transitional period between normal cognitive function and dementia in Parkinson's disease (PD). The expanded clinical subtypes are particularly relevant to this effort given the difference in presentation between dementias of different origins. Data suggest that MCI in those with Parkinson's is predictive of future dementia in much the same way that it is for individuals with MCI without co-morbid PD. Non-amnestic subtypes are particularly prevalent in Parkinson's disease and the presence of MCI in individuals with Parkinson's disease does substantially raise one's risk of developing dementia as compared to those with PD and normal cognition (Janvin *et al.*, 2006). In contrast to AD, conclusions about amnestic subtypes of PD MCI are more difficult to draw given the low prevalence of this subtype. Boyle *et al.* (2005) found that those individuals with MCI had higher levels of parkinsonian symptoms (though not Parkinson's disease) than those who were cognitively normal. Verghese *et al.* (2008) found greater gait abnormalities in those with aMCI as compared to those with non-amnestic MCI or controls while others have noted that non-amnestic subtypes have higher rates of gait dysfunction than amnestic MCI (Boyle *et al.*, 2005). Those with mMCI may present with more extra-pyramidal features than those with aMCI (Zanetti *et al.*, 2006).

VI. Daily Functioning and MCI

Embedded in the controversy surrounding the establishment of specific diagnostic criteria for MCI, there is much debate regarding whether impairment in everyday activities should be included as a criterion. In its initial conceptualization, MCI guidelines required that functional abilities remain intact (Petersen *et al.*, 1999) as this specific criterion helped distinguish MCI from dementia. However, as Farias *et al.* (2006) note, cognitive and functional deterioration clearly occurs over the course of MCI since such change eventually leads to conversion to dementia for many individuals with MCI (e.g., Bruscoli and Lovestone, 2004; Petersen *et al.*, 1999). In light of this accumulating evidence, an international working group proposed modified criteria for MCI, which includes "preserved basic activities of daily living" and "minimal impairment in complex instrumental functions" (Winblad *et al.*, 2004, p. 243). Similarly, participants of the International Psychogeriatric Association Expert Conference on MCI did not require intact ADL/instrumental activities of daily living (IADL) as a criterion but, instead, defined MCI as "a syndrome defined as cognitive decline greater than expected for an individual's age and education level but that does not interfere notably with activities of daily life" (Gauthier *et al.*, 2006, p. 1262).

Emerging information about the functional status of those with distinct clinical subtypes of MCI may shed additional light on the continuum of functional abilities spanning normal cognition to dementia. Supporting the notion that functional decline occurs on a continuum, several groups have reported that IADL decrements in MCI are intermediate to the subtle declines associated with normal aging and the frank impairments required for a dementia diagnosis (Farias *et al.*, 2006; Giovanetti *et al.*, 2008; Griffith *et al.*, 2003; Peres *et al.*, 2006).

Published reports comparing ADL and IADL performance between MCI individuals and their cognitively normal counterparts have demonstrated that those with MCI show greater IADL changes in areas including shopping, managing medications, and handling finances (Mariani *et al.*, 2008). In a study examining IADL performance across different MCI subtypes, Tam *et al.* (2007) reported that individuals with multiple domain MCI demonstrated impairment in IADL relative to both amnestic MCI participants and cognitively normal older adults. Corroborating this finding, Zanetti and colleagues (2006) demonstrated that individuals with multi-domain MCI performed more poorly on measures of ADL and IADL relative to amnestic MCI participants. In contrast, Farias and colleagues (2006) demonstrated that the MCI with memory impairment group showed somewhat more functional change relative to their MCI peers with no memory impairment and cognitively normal older adults. Similarly, Wadley and colleagues (2007) reported that amnestic, non-amnestic, and multiple domain MCI subgroups all demonstrated greater difficulty with IADL performance compared to the cognitively normal group. However, unlike the other two MCI groups, individuals with non-amnestic MCI did not differ from the cognitively normal participants in terms of IADL performance. Notably, at least one study (Boeve *et al.*, 2003) did not find differences between amnestic MCI individuals and cognitively normal older adults in terms of functional abilities. However, it should be noted that the MCI participants who were included in this study were characterized as MCI based, in part, on intact ADL.

Farias and colleagues (2006) administered self-report and informant-based versions of the Daily Function Questionnaire (DFQ) and calculated a difference score by subtracting patients' DFQ from informants' DFQ. DFQ difference scores, which indicate a lack of awareness of deficits, were greater in demented individuals relative to their cognitively normal counterparts and MCI subtype groups (i.e., MCI with memory impairment and MCI without memory impairment). However, the difference scores did not differ between MCI groups and cognitively normal older adults. Based on these findings, Farias and colleagues (2006) argued that individuals with MCI do not underestimate functional changes as is often the case for demented individuals.

What seems clear from the above discussion is that changes in ADL and IADL are not uncommon across the spectrum of MCI; although, it remains to be determined whether many of these changes would be conceptualized as frank "deficits" or "impairments." Comparing the ADL and IADL changes of MCI to overt dementia groups would be helpful in determining cutoff criteria for impairment, or administering performance-based ADL and IADL measures with normative reference standards would also help to delineate whether such changes represent impairment.

VII. Neuroimaging

A. Structural MRI

In determining the clinical viability of the various clinical subtypes, many would assert that different subtypes should have distinct neuropathology or different courses of change in brain integrity. Certainly, structural neuroimaging provides a non-invasive way to begin to examine brain changes associated with MCI, and there is emerging evidence to support distinct neuropathological profiles in clinical subtypes of MCI. Whitwell *et al.* (2007) found that those with amnestic presentations (single or multiple domain) had greater gray matter atrophy in medial and inferior temporal lobes compared to controls. Those with multi-domain amnestic MCI additionally showed loss in posterior temporal lobe, parietal association cortex, posterior cingulate, anterior insula, and the medial frontal lobe, a pattern of atrophy similar to that found in AD (Whitwell *et al.*, 2007). In support of this finding, Seo and colleagues (2007) reported that those diagnosed with single domain amnestic MCI showed cortical thinning in left medial temporal lobe (MTL) only, whereas those identified as multi-domain amnestic MCI showed cortical thinning in the left MTL, precuneus, and anterior and inferior basal temporal, insular, and temporal association cortices. The precuneus atrophy may be responsible for additional cognitive impairments present in the multi-domain MCI subtype and may suggest that the multi-domain presentations are a progression from single domain presentations since the areas of thinning noted in the multi-domain subtype encompassed all those in the single domain subtype and the extent of MTL atrophy was greater in the multi-domain versus the single domain subtype (Seo *et al.*, 2007).

In contrast, Becker and colleagues (2006) did not support the multi-domain subtype as the more advanced, transitional state between normal cognition and AD. They found that hippocampal volumes in those with multi-domain MCI were not statistically different from those of controls, but were significantly larger than both the amnestic MCI and AD groups (Becker *et al.*, 2006). Bell McGinty and colleagues (2005) found that the amnestic MCI group had greater volume loss in left entorhinal cortex and inferior parietal lobe as compared with multi-domain MCI. However, the multi-domain MCI group may exhibit their neuropathological changes in other areas, namely by smaller right inferior frontal gyrus, right middle temporal gyrus, and bilateral superior temporal gyrus as compared to amnestic MCI (Becker *et al.*, 2006).

Although the data are conflicting as to whether multi-domain subtypes necessarily have more extensive brain changes than single domain subtypes, current research does support the idea that different clinical subtypes of MCI have distinct neuropathology. Taken together, the available evidence suggests that those individuals with a more focal memory presentation have greater involvement of mesial temporal structures while those with more widespread deficits had greater involvement of association areas. Distinct MCI subtypes may represent different etiological paths to dementia, but the small sample sizes available in most of the imaging studies to date make conclusions tentative at best.

Other advanced imaging techniques hold promise to further clarify the nature and extent of brain changes associated with distinct clinical MCI subtypes though, to date, use of techniques such as functional magnetic resonance imaging (FMRI) and diffusion tensor imaging (DTI) has focused globally on MCI or on amnestic MCI, without significant investigation of non-amnestic subtypes. An overview of the use of these imaging techniques in MCI is provided, nonetheless, as this preliminary work is an essential framework for future examinations of clinical MCI subtypes.

B. Diffusion Tensor Imaging

A growing body of research suggests that white matter pathology contributes to age-related cognitive impairment and possibly potentiates the development of dementia (Raz and Rodrigue, 2006; Sullivan and Pfefferbaum, 2006). Although studies have generally shown white matter changes to be accelerated and more severe in AD (Pfefferbaum *et al.*, 2000; Rose *et al.*, 2006; Takahashi *et al.*, 2002), to date, few studies have employed DTI to examine early white matter changes in older adults with MCI with exceptionally limited focus on DTI-derived white matter changes in specific MCI clinical subtypes. Several studies have shown reduced fractional anisotropy (FA), a proxy of white matter integrity, in the posterior cingulum fibers, and this relationship seems to be stronger in the left versus right hemisphere (Fellgiebel *et al.*, 2005; Medina *et al.*, 2006; Rose *et al.*, 2006; Zhang *et al.*, 2007). Rose *et al.* (2006) demonstrated increased diffusivity in the entorhinal and parieto-occipital cortices, and decreased FA in the limbic parahippocampal white matter in patients with MCI. Moreover, Kantarci *et al.* (2005) was among the first to show that increased mean diffusivity of the hippocampus in amnestic MCI predicted future progression to dementia.

Several studies have shown decreased integrity in the posterior region of the corpus callosum (i.e., splenium) in those with MCI (Cho *et al.*, 2008; Delano-Wood *et al.*, 2007; Ukmar *et al.*, 2008), an area which is particularly sensitive to degenerative processes (Naggara *et al.*, 2006; Rose *et al.*, 2000; Takahashi *et al.*, 2002). Although some studies have shown changes in the frontal white matter of MCI patients (Bozzali *et al.*, 2002; Medina *et al.*, 2006; Naggara *et al.*, 2006), other studies have not identified any differences (Delano-Wood *et al.*, 2007; Head *et al.*, 2004; Medina *et al.*, 2006; Ukmar *et al.*, 2008). Although data are limited, results suggest a pattern of retrogenesis (Bartzokis, 2004), by which microstructural changes first occur in late-myelinating regions, spreading to early-myelinating regions only after the disease process has progressed beyond a particular threshold, which may initially manifest itself with the onset of MCI.

C. Functional MRI

Evidence to date indicates that functional brain decline precedes structural decline in prodromal dementia, including adults with MCI. Therefore, functional neuroimaging techniques may offer the unique ability to detect early functional brain changes in at-risk adults and identify the neurophysiological markers that best predict dementia conversion.

Given that AD neuropathology preferentially targets the MTL early in the course of the disease, thereby resulting in the hallmark episodic memory decline, and amnestic MCI is thought to represent prodromal AD, the majority of FMRI studies of MCI involve memory processing (particularly encoding) in amnestic samples. No known FMRI studies have been published focusing on other clinical subtypes of MCI. While several studies demonstrate increased blood oxygen level dependent (BOLD) response in the MTL (Dickerson *et al.*, 2004, 2005; Hamalainen *et al.*, 2007; Kircher *et al.*, 2007; Sperling, 2007), others report decreased MTL activity in MCI (Johnson *et al.*, 2006; Machulda *et al.*, 2003; Mandzia *et al.*, 2009). These discrepant findings have been interpreted as reflecting bimodal functional activity whereby less impaired MCI subjects show increased BOLD response in the hippocampus corresponding to a slight or moderate neuronal dysfunction, and more impaired MCI subjects demonstrate decreased BOLD response—similar to the levels observed in mild AD patients—as the cortical neuronal networks become more severely impaired with greater disease progression (Celone *et al.*, 2006; Dickerson *et al.*, 2004, 2005; Hamalainen *et al.*, 2007; Johnson *et al.*, 2006; Machulda *et al.*, 2003; Masdeu *et al.*, 2005; Petrella *et al.*, 2007a). However, this interpretation is primarily derived from cross-sectional studies and can only adequately be tested with longitudinal designs.

Few longitudinal FMRI studies of MCI have been reported. Although these studies are often limited by small sample sizes, they demonstrate promise for the use of FMRI to detect early AD. Those MCI patients who converted to AD showed a stronger relationship between brain activity in the left superior parietal lobe and the left precuneus during an angle discrimination task in the context of comparable performance (Vannini *et al.*, 2007). Similarly, despite equivalent memory performance, Dickerson *et al.* (2004) reported that MCI patients who subsequently declined during a 2.5-year follow-up period demonstrated increased right parahippocampal gyrus activity during picture encoding. In a more recent study, the same research group reported increased hippocampal activation predicted greater degree and rate of cognitive decline during a 6-year follow-up period, even after controlling for baseline level of impairment (Miller *et al.*, 2008).

Mandzia *et al.* (2009) reported that MTL activation during recognition was positively correlated with behavioral performance. However, unlike their healthy peers, MCI adults did not show a strong relationship between MTL activity during picture encoding and subsequent retrieval success, highlighting the complexity of the relationship between BOLD signal and effectiveness of encoding strategies. In contrast, Johnson *et al.* (2006) found reduced BOLD signal change in the right hippocampus during picture encoding and in the posterior cingulate during recognition of learned items in an amnestic MCI group despite comparable performance to their healthy peers. However, when activation corresponding only to successfully learned words was examined, an increase in hippocampal activity was seen, suggesting that an increase in MTL activity may support successful memory encoding (Kircher *et al.*, 2007). Similarly, a positive correlation between extent of parahippocampal and hippocampal activation and memory performance was found in MCI but, in a paradoxical fashion, greater clinical impairment, was also associated with recruitment of a larger region of the right parahippocampal gyrus during encoding (Dickerson *et al.*, 2004). Data from Johnson *et al.* (2004) provided further evidence for hippocampal dysfunction in MCI, suggesting that adults with MCI do not habituate to increasingly familiar items in the same manner as healthy older adults who show expected reductions in BOLD response to repeated items over time.

Despite the prevalence of studies examining medial temporal cortex function supporting memory, other cortical areas have also been implicated in MCI. For example, a reduction in functional activity in the posterior cingulate cortex (PCC) during recognition and episodic retrieval of previously learned line drawings (Johnson *et al.*, 2006) and object working memory (Yetkin *et al.*, 2006), but not during self-appraisal (Ries *et al.*, 2007), has implicated this region in the memory retrieval difficulty seen in amnestic MCI. The degradation of PCC functioning in MCI is not surprising given that PET metabolic alterations in the temporoparietal cortices and in the posterior cingulate have been reported in MCI and AD (Desgranges *et al.*, 1998; Matsuda, 2001; Reiman *et al.*, 1996) as well as in nondemented young and middle-aged adults at genetic risk for AD (Petrella *et al.*, 2007b; Reiman *et al.*, 1996, 2004, 2005; Wolf *et al.*, 2003). Similarly, dedifferentiation in the retrosplenial cortex during the retrieval of recent versus remote autobiographical memories and during episodic versus semantic memory retrieval has been reported in amnestic MCI (Poettrich *et al.*, 2009), further implicating the medial posterior cortex in MCI. Additionally, the neural substrates of visual working memory (Yetkin *et al.*, 2006), self-appraisal (Ries *et al.*, 2007), and emotional working memory (Dohnel *et al.*, 2008) in MCI have also been examined, and generally implicate a greater number of cortical regions. However, results are varied and highlight the need for greater attention to other cognitive processes in MCI in order to more fully understand changes in cortical functioning that may signal impending cognitive decline.

VIII. Treatment

One motivation to better understand the heterogeneous concept of MCI and the risk it imparts for future development of dementia is to provide early interventions that could halt or at least slow progression of symptoms. To date, unfortunately, there are no FDA-approved therapies for MCI. Further, aMCI has received all the attention with regard to treatment trials with no trials investigating other distinct clinical subtypes of MCI. Of the existing treatment

TABLE I
CLINICAL TRIALS IN aMCI

Agent	*N*	Duration	Endpoint	Outcome
Donepezil	269	24 weeks	Symptoms	Negative
Donepezil/Vitamin E	769	3 years	AD	Partially positive
Rofecoxib	1200	2–3 years	AD	Negative
Galantamine	995	2 years	CDR 1	Negative
	1062	2 years	CDR 1	Negative
Rivastigmine	1018	3–4 years	AD	Negative

AD, Alzheimer's disease; CDR, clinical dementia rating.

trials in MCI, most have used a "progression to AD" design with the focus on slowing cognitive decline and delaying conversion to AD. As a whole, the trials have been disappointing with one possible exception, the Alzheimer's Disease Cooperative Study (ADCS)-sponsored trial (Petersen *et al.*, 2005) (see Table I).

The ADCS-sponsored study of Vitamin E and donepezil for MCI involved 769 subjects at 69 centers in the US and Canada over 3 years. There were three treatment arms: Vitamin E 2000 IU/day, donepezil 10 mg/day, and placebo. The primary trial endpoint was conversion to AD. Although conversion to AD favored donepezil at 1 year, there were no differences among groups with regard to conversion to AD at 3 years. However, possession of the APOE ε4 allele was noted to be associated with a threefold greater risk of conversion from aMCI to dementia and, thus, clearly an important predictor of progression. When the authors looked at the progression to AD for APOE ε4 positive participants by treatment group, they found that the effect of donepezil was greater in ε4 positive individuals and persisted for 2 years. While neither of the two active arms reduced the risk of progressing to AD over the entire 36 months, donepezil reduced the risk of progression to AD for the first 12 months in all subjects and up to 24 months in those who were positive for the APOE ε4 allele. No treatment effect was noted for Vitamin E.

Other treatment trials have been less promising for halting conversion from MCI to dementia over time. A large trial of rivastigmine, an acetylcholinesterase inhibitor, was a double blind, placebo-controlled trial of 1018 patients that had many of the same features as the ADCS trial, but was conducted in 14 countries using multiple languages and translations of the neuropsychological instruments (Feldman *et al.*, 2007). At baseline, arms were not well matched with regard to frequency of APOE ε4 genotype, which was 46% in the placebo arm but only 37% in the rivastigmine arm. The study also had a lower conversion rate than expected and had to be extended to 4 years; over that time, 21.4% of placebo treated, but only 17.3% of rivastigmine treated, subjects progressed to AD.

Although rivastigmine was favored, the results were not statistically significant, and secondary assessments were also not significant. Investigation of the efficacy of another acetylcholinesterase inhibitor, galantamine, also failed to reveal a significant effect of galantamine on conversion to dementia in those with MCI in either of two trials (Winblad *et al.*, 2008).

Finally, another large randomized, placebo-controlled, double-blind study examined the ability of the COX 2 inhibitor, rofecoxib, to delay disease progression in 1457 aMCI subjects (Thal *et al.*, 2005). Once again, there was a lower than expected annual rate of conversion to AD. Conversion to AD actually favored placebo in this trial but the authors dismissed the significance of this finding because the secondary cognitive measures did not corroborate the primary outcome.

In hindsight, several important factors likely influenced the results of these studies, including, perhaps first and foremost, the variable rate of progression from aMCI to AD. Sources of this variability likely include subject heterogeneity, with regard to impairment level, culture, language, and APOE ε4 carrier status, in addition to even simple differences in implementation of enrollment criteria. MCI patients may show increased awareness of, or lower tolerability for, adverse events, resulting in higher discontinuation rates. Our current outcome measures may be insensitive; for example, the conversion design dichotomizes a continuous variable and most of the currently used efficacy measures follow an AD trial model of decline. Rather than decline, however, MCI patients may show improvement on cognitive measures, no matter which treatment group they are assigned to, because of at least some preservation in their ability to learn. Future MCI trials may benefit from less heterogeneous recruitment with stricter entry criteria and enriched populations, more sensitive cognitive and global outcome measures that reflect subtle impairments in complex activities, novel imaging outcomes, and longer trials.

IX. Conclusions

MCI remains a heterogeneous concept, though division of MCI into distinct clinical subtypes serves as a promising approach to better understanding MCI as a diagnostic entity and a risk factor for future cognitive decline. Evidence to date suggests that multi-domain amnestic presentations are more prevalent than either single domain amnestic or multi-domain non-amnestic presentations, though relatively little attention has been paid to the latter subtype. Converging neuropsychological, daily functioning, and neuroimaging data suggest that multiple domain presentations may place one at highest risk for future development of dementia. The current literature also supports that knowledge of subtypes of MCI informs the risk for future development of different types of dementia.

Though knowledge of MCI subtypes appears helpful in predicting risk of conversion to dementia, there remains a significant minority of individuals with MCI, particularly single domain subtype that may revert to normal cognition when followed over time. This instability in diagnosis as well as the varying prevalence rates, rates of conversion to dementia, and general oft-conflicting results in the literature, are likely due to ongoing challenges in operationalizing the diagnostic criteria for MCI. The methods for documenting objective neuropsychological impairment tend to be a particularly variable and ill-defined aspect of the MCI diagnostic process across studies (Portet *et al.*, 2006). There is little consensus about what neuropsychological tests (or how many) should be used to document objective cognitive impairment, what level of performance is considered cognitively impaired, how diagnostic criteria for different clinical subtypes of MCI are applied, what constitutes intact daily functioning, or about whether or not functional abilities should be included in the diagnostic decision-making regarding MCI. As this review highlights, the variable results in the current MCI literature clearly illustrate the importance of (a) understanding the criteria used to identify cognitive impairment in making the MCI diagnosis, (b) the value of using comprehensive neuropsychological assessment when diagnosing MCI subtypes, and (c) point to the need for further exploration of MCI subtypes, particularly non-amnestic presentations. Investigations and interventions targeting only amnestic subtypes are potentially missing a sizable number of individuals at risk.

Neuroimaging holds promise as a technique to better understand differences between distinct MCI subtypes although all of the above-mentioned methodological challenges together with small sample sizes and very limited attention paid to non-amnestic subtypes make drawing firm conclusions from the existing imaging literature challenging. To date, data do seem to support distinct neuropathology in the different clinical subtypes of MCI. However, there is still much overlap in structural imaging profiles and conflicting evidence making conclusions tentative at best. Advanced imaging techniques, such as DTI and functional MRI hold promise for detecting microstructural white matter damage or altered activation patterns in older adults prior to the manifestation of the full dementia syndrome. This early identification would identify the group in whom targeted therapies will likely have the greatest clinical impact (see Fagan *et al.*, 2005, for discussion). Overall, results from recent DTI studies indicate that white matter changes are evident in at-risk older adults and further validate the use of DTI to capture subtle, early white matter changes before significant atrophy is present. However, to date, very few studies have employed DTI in older adults with MCI, and even fewer have investigated the relationship between clinical subtype of MCI and white matter integrity.

Similarly, although FMRI techniques may prove to be instrumental in the early detection of AD, interpretation of current findings in MCI is complicated by various methodological differences between studies. In general, functional changes in the MTL and posterior medial cortex appear to signal cognitive decline and dementia conversion. However, discrepant results across studies may be due to differences in diagnostic classification of MCI adults. Specifically, although the majority of studies reviewed classified their patients as amnestic MCI, it is likely that the patient sample represented a more heterogeneous group that may reflect different underlying neural pathology. This highlights the need for future research aimed at integrating behavioral performance with measures of functional activity that directly compare different MCI subtypes with these sophisticated neuroimaging techniques.

Finally, results of intervention trials to halt the progression of MCI have generally been disappointing. Future trials are needed that address both amnestic and non-amnestic presentations, employ more stringent entry criteria, use more sensitive cognitive and global outcome measures that reflect subtle impairments in complex activities, and include novel imaging outcomes.

Acknowledgments

This work was supported by grants from the National Institutes of Health (K24 AG026431, R01 AG012674, and P50 AG05131), by Career Development Awards from the Department of Veterans Affairs, and by Investigator-Initiated and New Investigator Research Grants from the Alzheimer's Association. The authors gratefully acknowledge the assistance of staff, patients, and volunteers of the UCSD Alzheimer's Disease Research Center, and the UCSD Laboratory of Cognitive Imaging.

References

Alexopoulos, P., Grimmer, T., Perneczky, R., Domes, G., and Kurz, A. (2006). Progression to dementia in clinical subtypes of mild cognitive impairment. *Dement. Geriatr. Cogn. Disord.* **22,** 27–34.

Alladi, S., Arnold, R., Mitchell, J., Nestor, P. J., and Hodges, J. R. (2006). Mild cognitive impairment: Applicability of research criteria in a memory clinic and characterization of cognitive profile. *Psychol. Med.* **36,** 1–9.

Bartzokis, G. (2004). Age-related myelin breakdown: A developmental model of cognitive decline and Alzheimer's disease. *Neurobiol. Aging* **25,** 5–18.

Becker, J. T., Davis, S. W., Hayashi, K. M., Meltzer, C. C., Toga, A. W., Lopez, O. L., and Thompson, P. M. (2006). Three-dimensional patterns of hippocampal atrophy in mild cognitive impairment. *Arch. Neurol.* **63,** 97–101.

Bell-McGinty, S., Lopez, O. L., Meltzer, C. C., Scanlon, J. M., Whyte, E. M., DeKosky, S. T., *et al.* (2005). Differential cortical atrophy in subgroups of mild cognitive impairment. *Arch. Neurol.* **62,** 1393–1397.

Bickel, H., Mösch, E., Seigerschmidt, E., Siemen, M., and Förstl, H. (2006). Prevalence and persistence of mild cognitive impairment among elderly patients in general hospitals. *Dement. Geriatr. Cogn. Disord.* **21,** 242–250.

Boeve, B., McCormick, J., Smith, G., Ferman, T., Rummans, T., Carpenter, T., Ivnik, R., Kokmen, E., Tangalos, E., Edland, S., Knopman, D., and Petersen, R. (2003). Mild cognitive impairment in the oldest old. *Neurology* **60,** 477–480.

Boyle, P. A., Wilson, R. S., Aggarwal, N. T., Arvanitakis, Z., Kelly, J., Bienias, J. L., and Bennett, D. A. (2005). Parkinsonian signs in subjects with mild cognitive impairment. *Neurology* **65,** 1901–1906.

Bozzali, M., Falini, A., Franceschi, M., Cercignani, M., Zuffi, M., Scotti, G., Comi, G., and Filippi, M. (2002). White matter damage in Alzheimer's disease assessed in vivo using diffusion tensor magnetic resonance imaging. *J. Neurol. Neurosurg. Psychiatry* **72,** 742–746.

Bruscoli, M., and Lovestone, S. (2004). Is MCI really just early dementia? A systematic review of conversion studies. *Int. Psychogeriatr.* **16,** 12–40.

Busse, A., Bischkopf, J., Riedel-Heller, S. G., and Angermeyer, M. C. (2003). Subclassifications for mild cognitive impairment: Prevalence and predictive validity. *Psychol. Med.* **33,** 1029–1038.

Busse, A., Hensel, A., Guhne, U., Angermeyer, M. C., and Riedel-Heller, S. G. (2006). Mild cognitive impairment: Long-term course of four clinical subtypes. *Neurology* **67,** 2176–2185.

Celone, K. A., Calhoun, V. D., Dickerson, B. C., Atri, A., Chua, E. F., Miller, S. L., DePeau, K., Rentz, D. M., Selkoe, D. J., Blacker, D., Albert, M. S., and Sperling, R. A. (2006). Alterations in memory networks in mild cognitive impairment and Alzheimer's disease: An independent component analysis. *J. Neurosci.* **26,** 10222–10231.

Cho, H., Dong, W. Y., Shon, Y. M., Kim, B. S., Kim, Y. I., Choi, Y. B., Lee, K. S., Shim, Y. S., Yoon, B., Kim, W., and Ahn, K. J. (2008). Abnormal integrity of corticocortical tracts in mild cognitive impairment: A diffusion tensor imaging study. *J. Korean Med. Sci.* **23,** 477–483.

Das, S. K., Bose, P., Biswas, A., Dutt, A., Banerjee, T. K., Hazra, A., Raut, D. K., Chaudhuri, A., and Roy, T. (2007). An epidemiologic study of mild cognitive impairment in Kolkata, India. *Neurology* **68,** 2019–2026.

Debette, S., Bombois, S., Bruandet, A., Delbeuck, X., Lepoittevin, S., Delmaire, C., Leys, D., and Pasquier, F. (2007). Subcortical hyperintensities are associated with cognitive decline in patients with mild cognitive impairment. *Stroke* **38,** 2924–2930.

Delano-Wood, L., Jak, A. J., Schweinsburg, B., Wierenga, C., Horne, N., Salmon, D. P., Thal, L. J., Frank, L. R., and Bondi, M. W. (2007). Posterior white matter changes in MCI: Associations with cognition and stroke risk. *J. Int. Neuropscyhol. Assoc.* **13,** 81.

Desgranges, B., Baron, J. C., de la Sayette, V., Petit-Taboue, M. C., Benali, K., Landeau, B., Lechevalier, B., and Eustache, F. (1998). The neural substrates of memory systems impairment in Alzheimer's disease. A PET study of resting brain glucose utilization. *Brain* **121**(Pt 4), 611–631.

Di Carlo, A., Lamassa, M., Baldereschi, M., Inzitari, M., Scafato, E., Farchi, G., and Inzitari, D. (2007). CIND and MCI in the Italian elderly: Frequency, vascular risk factors, progression to dementia. *Neurology* **68,** 1909–1916.

Dickerson, B. C., Salat, D. H., Bates, J. F., Atiya, M., Killiany, R. J., Greve, D. N., Dale, A. M., Stern, C. E., Blacker, D., Albert, M. S., and Sperling, R. A. (2004). Medial temporal lobe function and structure in mild cognitive impairment. *Ann. Neurol.* **56,** 27–35.

Dickerson, B. C., Salat, D. H., Greve, D. N., Chua, E. F., Rand-Giovannetti, E., Rentz, D. M., Bertram, L., Mullin, K., Tanzi, R. E., Blacker, D., Albert, M. S., and Sperling, R. A. (2005). Increased hippocampal activation in mild cognitive impairment compared to normal aging and AD. *Neurology* **65,** 404–411.

Dohnel, K., Sommer, M., Ibach, B., Rothmayr, C., Meinhardt, J., and Hajak, G. (2008). Neural correlates of emotional working memory in patients with mild cognitive impairment. *Neuropsychologia* **46,** 37–48.

Fagan, A. M., Csernansky, C. A., Morris, J. C., and Holtzman, D. M. (2005). The search for antecedent biomarkers of Alzheimer's disease. *J. Alzheimer's Dis.* **8,** 347–358.

Farias, S. T., Mungas, D., Reed, B. R., Harvey, D., Cahn-Weiner, D., and DeCarli, C. (2006). MCI is associated with deficits in everyday functioning. *Alzheimer Dis. Assoc. Disord.* **20,** 217–223.

Feldman, H. H., Ferris, S., Winblad, B., Sfikas, N., Mancione, L., He, Y., Tekin, S., Burns, A., Cummings, J., del Ser, T., Inzitari, D., Orgogozo, J. M., *et al.* (2007). Effect of rivastigmine on delay to diagnosis of Alzheimer's disease from mild cognitive impairment: the InDDEx study. *Lancet* **6,** 501–512.

Fellgiebel, A., Muller, M. J., Wille, P., Dellani, P. R., Scheurich, A., Schmidt, L. G., and Stoeter, P (2005). Color-coded diffusion-tensor-imaging of posterior cingulate fiber tracts in mild cognitive impairment. *Neurobiol. Aging* **26,** 1193–1198.

Fischer, P., Jungwirth, S., Zehetmayer, S., Weissgram, S., Hoenigschnabl, S., Gelpi, E., Krampla, W., and Tragl, K. H. (2007). Conversion from subtypes of mild cognitive impairment to Alzheimer dementia. *Neurology* **68,** 288–291.

Gabryelewicz, T., Styczynska, M., Luczywek, E., Barczak, A., Pfeffer, A., Androsiuk, W., Chodakowska-Zebrowska, M., Wasiak, B., Peplonska, B., and Barcikowska, M. (2007). The rate of conversion of mild cognitive impairment to dementia: Predictive role of depression. *Int. J. Geriatr. Psychiatry* **22,** 563–567.

Gauthier, S., Reisberg, B., Zaudig, M., Petersen, R. C., Ritchie, K., Broich, K., Belleville, S., Brodaty, H., Bennett, D., Chertkow, H., Cummings, J. L., de Leon, M., *et al.* (2006). Mild cognitive impairment. *The Lancet* **367,** 1262–1270.

Giovanetti, T., Bettcher, B. M., Brennan, L., Libon, D. J., Burke, M., Duey, K., Nieves, C., and Wambach, D. (2008). Characterization of everyday functioning in mild cognitive impairment: A direct assessment approach. *Dement. Geriatr. Cogn. Disord.* **25,** 359–365.

Griffith, H. R., Belue, B. S., Sicola, A., Krzywanski, S., Zamrini, E., Harrell, L., and Marson, D. C. (2003). Impaired financial abilities in mild cognitive impairment: A direct assessment approach. *Neurology* **60,** 449–457.

Hamalainen, A., Pihlajamaki, M., Tanila, H., Hanninen, T., Niskanen, E., Tervo, S., Karjalainen, P. A., Vanninen, R. L., and Soininen, H. (2007). Increased fMRI responses during encoding in mild cognitive impairment. *Neurobiol. Aging* **28,** 1889–1903.

Head, D., Buckner, R. L., Shimony, J. S., Williams, L. E., Akbudak, E., Conturo, E., McAvoy, M., Morris, J. C., and Snyder, A. Z. (2004). Differential vulnerability of anterior white matter in nondemented aging with minimal acceleration in dementia of the Alzheimer's type: Evidence form diffusion tensor imaging. *Cerebral. Cortex* **14,** 410–423.

Hort, J., Laczo, J., Vyhnalek, M., Bojar, M., Bures, J., and Vlcek, K. (2007). Spatial navigation deficit in amnesic mild cognitive impairment. *Proc. Natl. Acad. Sci. USA* **104,** 4042–4047.

Jak, A. J., Corey-Bloom, J., and Bondi, M. W. (2007). Diagnostic characterization of MCI subtypes in a naturalistic sample. *J. Int. Neuropscyhol. Assoc.* **13,** 4.

Janvin, C. C., Larsen, J. P., Aarsland, D., and Hugdahl, K. (2006). Subtypes of mild cognitive impairment in Parkinson's disease: Progression to dementia. *Mov. Disord.* **21,** 1343–1349.

Johnson, S. C., Baxter, L. C., Susskind-Wilder, L., Connor, D. J., Sabbagh, M. N., and Caselli, R. J. (2004). Hippocampal adaptation to face repetition in healthy elderly and mild cognitive impairment. *Neuropsychologia* **42,** 980–989.

Johnson, S. C., Schmitz, T. W., Moritz, C. H., Meyerand, M. E., Rowley, H. A., Alexander, A. L., Hansen, K. W., Gleason, C. E., Carlsson, C. M., Ries, M. L., Asthana, S., Chen, K., *et al.* (2006). Activation of brain regions vulnerable to Alzheimer's disease: The effect of mild cognitive impairment. *Neurobiol. Aging* **27,** 1604–1612.

Jungwirth, S., Weissgram, S., Zehetmayer, S., Tragl, K. H., and Fischer, P. (2005). VITA: Subtypes of mild cognitive impairment in a community-based cohort at the age of 75 years. *Int. J. Geriatr. Psychiatry* **20,** 452–458.

Kantarci, K., Petersen, R. C., Boeve, B. F., Knopman, D. S., Weigand, S. D., O'Brien, P. C., *et al.* (2005). DWI predicts future progression to Alzheimer disease in amnestic mild cognitive impairment. *Neurology* **64,** 902–904.

Kircher, T. T., Weis, S., Freymann, K., Erb, M., Jessen, F., Grodd, W., Heun, R., and Leube, D. T. (2007). Hippocampal activation in patients with mild cognitive impairment is necessary for successful memory encoding. *J. Neurol. Neurosurg. Psychiatry* **78,** 812–818.

Kryscio, R. J., Schmitt, F. A., Salazar, J. C., Mendiondo, M. S., and Markesbery, W. R. (2006). Risk factors for transitions from normal to mild cognitive impairment and dementia. *Neurology* **66,** 828–832.

Lopez, O. L., Jagust, W. J., DeKosky, S. T., Becker, J. T., Fitzpatrick, A., Dulberg, C., Breitner, J., Lyketsos, C., Jones, B., Kawas, C., Carlson, M., and Kuller, L. H. (2003). Prevalence and classification of mild cognitive impairment in the cardiovascular health study cognition study: Part 1. *Arch. Neurol.* **60,** 1385–1389.

Lopez, O. L., Becker, J. T., Jagust, W. J., Fitzpatrick, A., Carlson, M. C., DeKosky, S. T., Breitner, J., Lyketsos, C. G., Jones, B., Kawas, C., and Kuller, L. H. (2006). Neuropsychological characteristics of mild cognitive impairment subgroups. *J. Neurol. Neurosurg. Psychiatry* **77,** 159–165.

Machulda, M. M., Ward, H. A., Borowski, B., Gunter, J. L., Cha, R. H., O'Brien, P. C., Petersen, R. C., Boeve, B. F., Knopman, D., Tang-Wai, D. F., Ivnik, R. J., Smith, G. E., *et al.* (2003). Comparison of memory fMRI response among normal, MCI, and Alzheimer's patients. *Neurology* **61,** 500–506.

Maioli, F., Coveri, M., Pagni, P., Chiandetti, C., Marchetti, C., Ciarrocchi, R., Ruggero, C., Nativio, V., Onesti, A., D'Anastasio, C., and Pedone, V. (2007). Conversion of mild cognitive impairment to dementia in elderly subjects: A preliminary study in memory and cognitive disorder unit. *Arch. Gerontol. Geriatr.* **44**(Suppl. 1), 233–241.

Mandzia, J. L., McAndrews, M. P., Grady, C. L., Graham, S. J., and Black, S. E. (2009). Neural correlates of incidental memory in mild cognitive impairment: An fMRI study. *Neurobiol. Aging* **30,** 717–730.

Manly, J. J., Bell-McGinty, S., Tang, M., Schupf, N., Stern, Y., and Mayeux, R. (2005). Implementing diagnostic criteria and estimating frequency of mild cognitive impairment in an urban community. *Arch. Neurol.* **62,** 1739–1746.

Mariani, E., Monastero, R., Ercolani, S., Mangialasche, F., Caputo, M., Feliziani, F. T., Vitale, D. F., Senin, U., and Mecocci, P. (2007). Vascular risk factors in mild cognitive impairment subytpes. *Dement. Geriatr. Cogn. Disord.* **24,** 448–456.

Mariani, E., Monastero, R., Ercolani, S., Rinaldi, P., Mangialasche, F., Costanzi, E., Vitale, D. F., Senin, U., and Mecocci, P. (2008). Influence of comorbidity and cognitive status on instrumental activities of daily living in amnestic mild cognitive impairment: Results from the ReGAl project. *Int. J. Geriatr. Psychiatry* **23,** 523–530.

Masdeu, J. C., Zubieta, J. L., and Arbizu, J. (2005). Neuroimaging as a marker of the onset and progression of Alzheimer's disease. *J. Neurol. Sci.* **236,** 55–64.

Matsuda, H. (2001). Cerebral blood flow and metabolic abnormalities in Alzheimer's disease. *Ann. Nucl. Med.* **15,** 85–92.

Medina, D., deToledo-Morrell, L., Urresta, F., Gabrieli, J. D. E., Moseley, M., Fleischman, D., Bennett, D. A., Leurgans, S., Turner, D. A., and Stebbins, G. T. (2006). White matter changes in mild cognitive impairment and AD: A diffusion tensor imaging study. *Neurobiol. Aging* **27,** 663–672.

Miller, S. L., Fenstermacher, E., Bates, J., Blacker, D., Sperling, R. A., and Dickerson, B. C. (2008). Hippocampal activation in adults with mild cognitive impairment predicts subsequent cognitive decline. *J. Neurol. Neurosurg. Psychiatry* **79,** 630–635.

Naggara, O., Oppenheim, C., Rieu, D., Raoux, N., Rodrigo, S., Barba, G. D., and Meder, J.-F. (2006). Diffusion tensor imaging in early Alzheimer's disease. *Psychiatry Res.* **146,** 663–672.

Palmer, K., Berger, A. K., Monastero, B., Winblad, B., Backman, L., and Fratiglioni, L. (2007). Predictors of progression from mild cognitive impairment to Alzheimer disease. *Neurology* **68,** 1596–1602.

Palmer, K., Backman, L., Winblad, B., and Fratiglioni, L. (2008). Mild cognitive impairment in the general population: Occurrence and progression to Alzheimer disease. *Am. J. Geriatr. Psychiatry* **16,** 603–611.

Peres, K., Chrysostome, V., Fabrigoule, C., Orgogozo, J. M., Dartigues, J. F., and Barberger-Gateau, P. (2006). Restriction in complex activities of daily living in MCI: Impact on outcome. *Neurology* **67,** 461–466.

Petersen, R. C. (2004). Mild cognitive impairment as a diagnostic entity. *J. Int. Med.* **256,** 183–194.

Petersen, R. C., and Morris, J. C. (2005). Mild cognitive impairment as a clinical entity and treatment target. *Arch. Neurol.* **62,** 1160–1163.

Petersen, R. C., Smith, G. E., Waring, S. C., Ivnik, R. J., Tangalos, E. G., and Kokmen, E. (1999). Mild cognitive impairment: Clinical characterization and outcome. *Arch. Neurol.* **56,** 303–308.

Petersen, R. C., Doody, R. S., Kurz, A., Mohs, R. C., Morris, J. C., Rabins, P. V., Ritchie, K., Rossor, M. N., Thal, L. J., and Winblad, B. (2001). Current concepts in mild cognitive impairment. *Arch. Neurol.* **58,** 1985–1992.

Petersen, R. C., Thomas, R. G., Grundman, M., Bennett, D., Doody, R., Ferris, S., Galasko, D., Jin, S., Kaye, J., Levey, A., Pfeiffer, E., Sano, M., *et al.* (2005). Vitamin E and donepezil for the treatment of mild cognitive impairment. *N. Engl. J. Med.* **52,** 2379–2388.

Petrella, J. R., Prince, S. E., Wang, L., Hellegers, C., and Doraiswamy, P. M. (2007a). Prognostic value of posteromedial cortex deactivation in mild cognitive impairment. *PLoS ONE* **2,** e1104.

Petrella, J. R., Wang, L., Krishnan, S., Slavin, M. J., Prince, S. E., Tran, T. T., and Doraiswamy, P. M. (2007b). Cortical deactivation in mild cognitive impairment: High-field-strength functional MR imaging. *Radiology* **245,** 224–235.

Pfefferbaum, A., Sullivan, E. V., Hedehus, M., Lim, K. O., Adalsteinsson, E., and Moseley, M. (2000). Age-related decline in brain white matter anisotropy measured with spatially corrected echo-planar diffusion tensor imaging. *Magn. Reson. Med.* **44,** 259–268.

Poettrich, K., Weiss, P. H., Werner, A., Lux, S., Donix, M., Gerber, J., von Kummer, R., Fink, G. R., and Holthoff, V. A. (2009). Altered neural network supporting declarative long-term memory in mild cognitive impairment. *Neurobiol. Aging* **30,** 284–298.

Portet, F., Ousset, P. J., Visser, P. J., Frisoni, G. B., Nobili, F., Scheltens, P., Vellas, B., and Touchon, J. the MCI Working Group of the European Consortium on Alzheimer's Disease (EADC) (2006). Mild cognitive impairment (MCI) in medical practice: A critical review of the concept and new diagnostic procedure. Report of the MCI Working Group of the European Consortium on Alzheimer's Disease. *J. Neurol. Neurosurg. Psychiatry* **77,** 714–718.

Rasquin, S. M. C., Lodder, J., Visser, P. J., Lousberg, R., and Verhey, F. R. J. (2005). Predictive accuracy of MCI subtypes for Alzheimer's disease and vascular dementia in subjects with mild cognitive impairment: A 2-year follow-up study. *Dement. Geriatr. Cogn. Disord.* **19,** 113–119.

Ravaglia, G., Forti, P., Maioli, F., Martelli, M., Servadei, L., Brunetti, N., Pantieri, G., and Mariani, E. (2006). Conversion of mild cognitive impairment to dementia: Predictive role of mild cognitive impairment subtypes and vascular risk factors. *Dement. Geriatr. Cogn. Disord.* **21,** 51–58.

Raz, N., and Rodrigue, K. M. (2006). Differential aging of the brain: Patterns, cognitive correlates and modifiers. *Neurosci. Biobehav. Rev.* **30,** 730–748.

Reiman, E. M., Caselli, R. J., Yun, L. S., Chen, K., Bandy, D., Minoshima, S., Thibodeau, S. N., and Osborne, D. (1996). Preclinical evidence of Alzheimer's disease in persons homozygous for the epsilon 4 allele for apolipoprotein E. *N. Engl. J. Med.* **334,** 752–758.

Reiman, E. M., Chen, K., Alexander, G. E., Caselli, R. J., Bandy, D., Osborne, D., Saunders, A. M., and Hardy, J. (2004). Functional brain abnormalities in young adults at genetic risk for late-onset Alzheimer's dementia. *Proc. Natl. Acad. Sci. USA* **101,** 284–289.

Reiman, E. M., Chen, K., Alexander, G. E., Caselli, R. J., Bandy, D., Osborne, D., Saunders, A. M., and Hardy, J. (2005). Correlations between apolipoprotein E epsilon4 gene dose and brain-imaging measurements of regional hypometabolism. *Proc. Natl. Acad. Sci. USA* **102,** 8299–8302.

Ries, M. L., Jabbar, B. M., Schmitz, T. W., Trivedi, M. A., Gleason, C. E., Carlsson, C. M., Rowley, H. A., Asthana, S., and Johnson, S. C. (2007). Anosognosia in mild cognitive impairment: Relationship to activation of cortical midline structures involved in self-appraisal. *J. Int. Neuropsychol. Soc.* **13,** 450–461.

Rose, S. E., Chen, F., Chalk, J. B., Zelaya, F. O., Strugnell, W. E., Benson, M., Semple, J., and Doddrell, D. M. (2000). Loss of connectivity in Alzheimer's disease: An evaluation of white matter tract integrity with colour coded MR diffusion tensor imaging. *J. Neurol. Neurosurg. Psychiatry* **69,** 528–530.

Rose, S. E., McMahon, K. L., Janke, A. L., O'Dowd, B., de Zubicaray, G., Strudwick, M. W., and Chalk, J. B. (2006). Diffusion indices on magnetic resonance imaging and neuropsychological performance in amnestic mild cognitive impairment. *J. Neurol. Neurosurg. Psychaitry* **77,** 1122–1128.

Rountree, S. D., Waring, S. C., Chan, W. C., Lupo, P. J., Darby, E. J., and Doody, R. S. (2007). Importance of subtle amnestic and nonamnestic deficits in mild cognitive impairment: Prognosis and conversion to dementia. *Dement. Geriatr. Cogn. Disord.* **24,** 476–482.

Rozzini, L., Chilovi, B. V., Conti, M., Delrio, I., Borroni, B., Trabucchi, M., and Padovani, A. (2008). Neuropsychiatric symptoms in amnestic and nonamnestic mild cognitive impairment. *Dement. Geriatr. Cogn. Disord.* **25,** 32–36.

Seo, S. W., Im, K., Lee, J., Kim, Y., Kim, S. T., Kim, S. Y., Yang, D. W., Kim, S. I., Cho, Y. S., and Na, D. L. (2007). Cortical thickness in single- versus multiple-domain amnestic mild cognitive impairment. *NeuroImage* **36,** 289–297.

Smith, G. E., and Ivnik, R. J. (2003). Normative neuropsychology. *In* "Mild Cognitive Impairment: Aging to Alzheimer's Disease" (R. C. Petersen, Ed.), pp. 63–88. Oxford University Press, New York.

Sperling, R. (2007). Functional MRI studies of associative encoding in normal aging, mild cognitive impairment, and Alzheimer's disease. *Ann. NY Acad. Sci.* **1097,** 146–155.

Sullivan, E. V., and Pfefferbaum, A. (2006). Diffusion tensor imaging and aging. *Neurosci. Biobehav. Rev.* **30,** 749–761.

Tabert, M. H., Manly, J. J., Liu, X., Pelton, G. H., Rosenblum, S., Jacobs, M., Zamora, D., Goodkind, M., Bell, K., Stern, Y., and Devanand, D. P. (2006). Neuropsychological prediction of conversion to alzheimer disease in patients with mild cognitive impairment. *Arch. Gen. Psychiatry* **63,** 916–924.

Takahashi, S., Yonezawa, H., Takahashi, J., Kudo, M., Inoue, T., and Tohgi, H. (2002). Selective reduction of diffusion anisotropy in white matter of Alzheimer disease brains measured by 3.0 Tesla magnetic resonance imaging. *Neurosci. Lett.* **332,** 45–48.

Tam, C., Lam, L., Chiu, H. F. K., and Lui, V. W. C. (2007). Characteristic Profiles of Instrumental Activities of Daily Living in Chinese Older Persons with Mild Cognitive Impairment. *Am. J. Alzheimer's Dis. Other Dement.* **22,** 211–217.

Teng, E., Lu, P. H., and Cummings, J. L. (2007). Deficits in facial emotion processing in mild cognitive impairment. *Dement. Geriatr. Cogn. Disord.* **23,** 271–279.

Thal, L. J., Ferris, S. H., Kirby, L., Block, G. A., Lines, C. R., Yuen, E., Assaid, C., Nessly, M. L., Norman, B. A., Baranak, C. C., and Reines, S. A. (2005). A randomized, double-blind, study of rofecoxib in patients with mild cognitive impairment. *Neuropsychopharmacology* **30,** 1204–1215.

Ukmar, M., Makuc, E., Onor, M. L., Garbin, G., Trevisiol, M., and Cova, M. A. (2008). Evaluation of white matter damage in patients with Alzheimer's disease and in patients with mild cognitive impairment by using diffusion tensor imaging. *Radiol. Med.* **113,** 915–922.

Vannini, P., Almkvist, O., Dierks, T., Lehmann, C., and Wahlund, L. O. (2007). Reduced neuronal efficacy in progressive mild cognitive impairment: A prospective fMRI study on visuospatial processing. *Psychiatry Res.* **156,** 43–57.

Verghese, J., Robbins, M., Holtzer, R., Zimmerman, M. E., Wang, C., Xue, X., and Lipton, R. B. (2008). Gait dysfunction in mild cognitive impairment syndromes. *J. Am. Geriatr. Soc.* **56,** 1244–1251.

Wadley, V. G., Crowe, M., Marsiske, M., Cook, S. E., Unverzagt, F. W., Rosenberg, A. L., *et al.* (2007). Changes in Everyday Function in Individuals with Psychometrically Defined Mild Cognitive Impairment in the Advanced Cognitive Training for Independent and Vital Elderly Study. *J. Am. Geriatr. Soc.* **55,** 1192–1198.

Westervelt, H. J., Bruce, J. M., Coon, W. G., and Tremont, G. (2008). Odor identification in mild cognitive impairment subtypes. *J. Clin. Exp. Neuropsychol.* **30,** 151–156.

Whitwell, J. L., Przybelski, S. A., Weigand, S. D., Knopman, D., Boeve, B. F., Petersen, R. C., and Jack, C. R. (2007). 3D maps from multiple MRI illustrate changing atrophy patterns as subjects progress from mild cognitive impairment to Alzheimer's disease. *Brain* **130,** 1777–1786.

Winblad, B., Palmer, K., Kivipelto, M., Jelic, V., Fratiglioni, L., Wahlund, L.-O., Nordberg, A., Backman, L., Albert, M., Almkvist, O., Arai, H., Basun, H., *et al.* (2004). Mild cognitive impairment—Beyond controversies, towards a consensus: Report of the International Working Group on Mild Cognitive Impairment. *J. Int. Med.* **256,** 240–246.

Winblad, B., Gauthier, S., Scinto, L., Feldman, H., Wilcock, G. K., Truyen, L., Mayorga, A. J., Wang, D., Brashear, H. R., and Nye, J. S. (2008). Safety and efficacy of galantamine in subjects with mild cognitive impairment. *Neurology* **70,** 2024–2035.

Wolf, H., Jelic, V., Gertz, H. J., Nordberg, A., Julin, P., and Wahlund, L. O. (2003). A critical discussion of the role of neuroimaging in mild cognitive impairment. *Acta Neurol. Scand. Suppl.* **179,** 52–76.

Yaffe, K., Petersen, R. C., Lindquist, K., Kramer, J., and Miller, B. (2006). Subtype of mild cognitive impairment and progression to dementia and death. *Dement. Geriatr. Cogn. Disord.* **22,** 312–319.

Yetkin, F. Z., Rosenberg, R. N., Weiner, M. F., Purdy, P. D., and Cullum, C. M. (2006). FMRI of working memory in patients with mild cognitive impairment and probable Alzheimer's disease. *Eur. Radiol.* **16,** 193–206.

Zanetti, M., Ballabio, C., Abbate, C., Cutaia, C., Vergani, C., and Bergamaschini, L. (2006). Mild cognitive impairment subtypes and vascular dementia in community-dwelling elderly people: A 3-year follow-up study. *J. Am. Geriatr. Soc.* **54,** 580–586.

Zhang, Y., Schuff, N., Jahng, G. H., Bayne, W., Mori, S., Schad, L., Mueller, S., Du, A. T., Kramer, J. H., Yaffe, K., Chui, H. C., Jagust, W. J., *et al.* (2007). Diffusion tensor imaging of cingulum fibers in mild cognitive impairment and Alzheimer disease. *Neurology* **68,** 13–19.

PROTON MAGNETIC RESONANCE SPECTROSCOPY IN DEMENTIAS AND MILD COGNITIVE IMPAIRMENT

H. Randall Griffith,[*,†,‡] Christopher C. Stewart,[‡] and Jan A. den Hollander[§]

[*]Department of Neurology, University of Alabama at Birmingham, Birmingham, Alabama 35233, USA

[†]Alzheimer's Disease Research Center, University of Alabama at Birmingham, Birmingham, Alabama 35233, USA

[‡]Department of Psychology, University of Alabama at Birmingham, Birmingham, Alabama 35233, USA

[§]Department of Medicine (Cardiology), University of Alabama at Birmingham, Birmingham, Alabama 35233, USA

With the anticipated increase in dementias due to the aging demographic of industrialized nations, biomarkers for neurodegenerative diseases are increasingly important as new therapies are being developed for clinical trials. Proton MR spectroscopy (^{1}H MRS) appears poised to be a viable means of tracking brain metabolic changes due to neurodegenerative diseases and potentially as a biomarker for treatment effects in clinical therapeutic trials. This review highlights the body of literature investigating brain metabolic abnormalities in Alzheimer's disease, amnestic mild cognitive impairment, frontotemporal dementia, vascular dementia, Lewy body dementia, and Parkinson's disease dementia. In particular, the review addresses the viability of ^{1}H MRS to discriminate among dementias, to measure disease progression, and to measure the effects of pharmacological treatments. While findings to date are encouraging, more study is needed in longitudinal patterns of brain metabolic changes, correspondence with changes in clinical markers of disease progression, and sensitivity of ^{1}H MRS measures to

INTERNATIONAL REVIEW OF NEUROBIOLOGY, VOL. 84
DOI: 10.1016/S0074-7742(09)00406-1

0074-7742/09 $35.00

treatment effects. Such developments will hopefully benefit the search for effective treatments of dementias in the twenty-first century.

I. Introduction

The aging demographic of persons in the United States and other industrialized nations creates the prospect for a future major healthcare crisis from providing care to patients with dementia worldwide. Clearly, there is a great need for better treatments to emerge for these neurodegenerative illnesses within the next decade. A prime issue in clinical investigation of treating neurodegenerative diseases is the need for better markers of the disease process that can be used as measures of treatment efficacy (Mueller *et al.*, 2006). The current review focuses on proton magnetic resonance spectroscopy (^{1}H MRS) as a marker of metabolic brain changes in dementias and the potential for use as a biomarker in clinical trials. Findings are discussed from studies of Alzheimer's disease (AD), amnestic mild cognitive impairment (MCI), frontotemporal dementia (FTD), vascular dementia (VAD), Lewy body dementia (LBD), and Parkinson's disease dementia (PDD).

II. Metabolites in ^{1}H MRS

A comprehensive review of the development and technology of ^{1}H MRS is beyond the scope of the current review. Interested readers are referred to Ross and Bluml (2001). Although phosphorus (^{31}P), carbon (^{13}C), and fluorine (^{19}F) are all forms of MRS (Mueller *et al.*, 2006), ^{1}H MRS is the most frequently encountered subtype in the dementia literature due to its superior MR signal sensitivity, good spatial resolution, and its widespread availability (Jones and Waldman, 2004; Ross and Bluml, 2001). This review will thus focus exclusively on (Mueller *et al.*, 2006) ^{1}H MRS, and the metabolites commonly assessed in these studies. The following metabolites are mainly of interest in dementias (see Fig. 1).

N-acetyleaspartate. *N*-acetyleaspartate (NAA), which resonates at 2.02 parts per million (ppm), represents the largest proton metabolic concentration in the human brain after H_2O (Kwock, 1998; Valenzuela and Sachdev, 2001). NAA exists primarily within neurons, axons, and their processes, and has been implicated in several neuronal processes, including lipid and protein synthesis, mitochondrial functioning, and osmoregulation (Jones and Waldman, 2004). NAA is presumed to represent neuronal density or integrity based upon comparison of in vivo and in vitro studies (Cheng *et al.*, 2002; Valenzuela and Sachdev, 2001) and the prevalence of reduced NAA levels within a number of neuropathological conditions (Chen *et al.*, 2000). NAA may also serve as a marker of axonal metabolic fitness

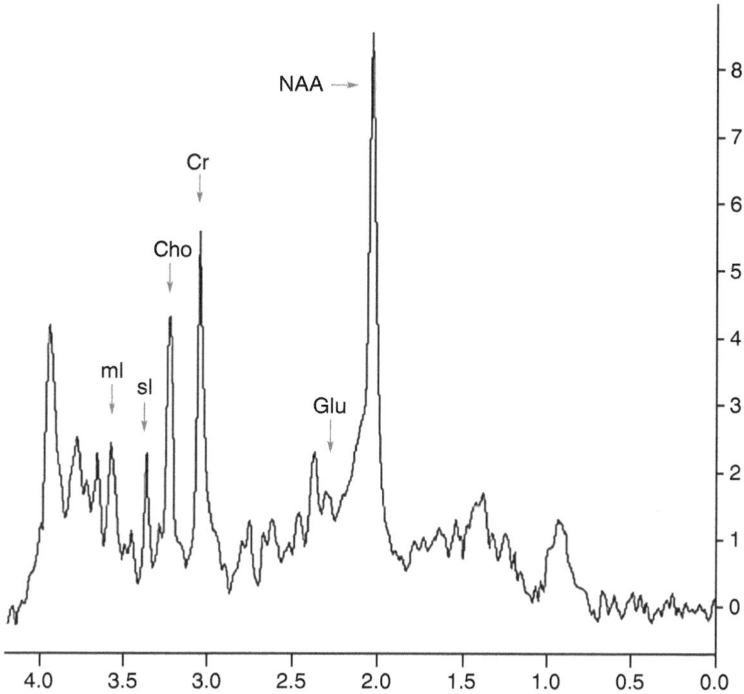

FIG. 1. Example spectra obtained from posterior cingulate gyrus at 3 T. mI, myo-Inositol; sI, scyllo-Inositol; Cho, Choline; Cr, Creatine; Glu, glutamate; NAA, *N*-acetyleaspartate.

(Valenzuela and Sachdev, 2001), as white and gray matter have comparable NAA levels and white matter NAA is reduced in white matter diseases (e.g., multiple sclerosis). Normalization of NAA levels in neurological disorders has been noted (e.g., following stroke); however, it remains unclear whether NAA "recovery" represents actual neuronal "recovery" or "regeneration" (Mueller *et al.*, 2006).

Choline. The Choline (Cho) peak (3.2 ppm) represents a combination of several choline-containing compounds, including free Cho, phosphorylcholine and glycerophosphorylcholine, and to a small extent acetylcholine (Firbank *et al.*, 2002; Valenzuela and Sachdev, 2001). Free Cho acts as a precursor to acetylcholine, while glycerophosphorylcholine is a product of breakdown of membrane phosphatidylcholine and acts as an osmoregulator (Firbank *et al.*, 2002). Cho is slightly more concentrated in white versus gray matter (Mueller *et al.*, 2006) and is likely present in glia. The Cho peak is often viewed as a marker of membrane turnover or inflammation in ^{1}H MRS studies (Mueller *et al.*, 2006).

Creatine. The Creatine (Cr) peak (3.0 ppm) is composed of creatine and phosphocreatine. Together, these metabolites buffer the energy use and energy storage of cells (Valenzuela and Sachdev, 2001). Cr is thought to exist primarily in glia and in neurons to a lesser extent (Firbank *et al.*, 2002) and is found in greater concentrations in human gray versus white matter (Mueller *et al.*, 2006). Cr concentration is often used as an internal standard because it is considered to be relatively stable, showing slow increases with age (Firbank *et al.*, 2002; Valenzuela and Sachdev, 2001; however see Rosen and Lenkinski, 2007).

myo-Inositol. The myo-Inositol (mI) peak (3.56 ppm) represents the presence of myo-Inositol, a sugar alcohol that is similar in structure to glucose, and myo-Inositol phosphate (Valenzuela and Sachdev, 2001). These metabolites are located primarily in glia, where they are speculated to function as osmoregulators, intracellular messengers, and detoxifying agents (Valenzuela and Sachdev, 2001). mI is highly concentrated in gliosing lesions (Firbank *et al.*, 2002) and is one of the products of myelin breakdown (Rosen and Lenkinski, 2007). Thus, the mI peak is thought to represent an increase in glia density or an increase in glia size and associated brain inflammation (Brand *et al.*, 1993; Rosen and Lenkinski, 2007; Strange *et al.*, 1994; Valenzuela and Sachdev, 2001).

Scyllo-Inositol. The peak resonating at 3.342 ppm is thought to represent the presence of scyllo-Inositol (sI), a product of mI metabolism (McLaurin *et al.*, 2000). The neurobiological function of sI as differentiated from mI is unclear.

Glutamate-Glutamine. The GLX peak (2.1-2.4 ppm) represents both glutamate (Glu) and glutamine (Gln), which cannot be individually distinguished on 1.5 T (Mueller *et al.*, 2006) ^{1}H MRS but is more reliably distinguishable at 3 T (Kantarci *et al.*, 2003; Schubert *et al.*, 2004). Glu acts as the major excitatory neurotransmitter in the brain, and Gln is thought to play a role in the regulation and detoxification of glutamate (Valenzuela and Sachdev, 2001). Both Glu and Gln have been implicated in neuronal function, metabolism, and plasticity (Antuono *et al.*, 2001). Cycling between Glu and Gln is believed to account for approximately 80% of cerebral metabolism (Magistretti and Pellerin, 1999). Additionally, extracellular glutamate is excitotoxic and is now a target of treatment in neurodegenerative dementias (Riederer and Hoyer, 2006). However, the glutamate signal in brain (Mueller *et al.*, 2006) ^{1}H MRS is thought to primarily represent the intracellular compartment (Kickler *et al.*, 2007).

III. ^{1}H MRS Findings in Dementias

A. ^{1}H MRS Findings in Alzheimer's Disease

Dementia due to AD is the most common dementia affecting older adults (McMurtray *et al.*, 2006). The neuropathology of AD consists of amyloid plaques and neurofibrillary tangles that present in a characteristic pattern with early involvement of the medial temporal lobes and hippocampus, with subsequent spread to paralimbic and association cortical regions (Braak and Braak, 1991).

Definitive diagnosis of AD requires microscopic neuropathological examination and is most commonly made postmortem (Cummings *et al.*, 1998). Thus, the presence and pattern of cortical metabolic abnormalities are of potential clinical value in identifying the underlying neuropathology of AD and discriminating this disease from other dementias (Kantarci *et al.*, 2008).

Reduction in NAA is the most frequent ^{1}H MRS finding in AD (Adalsteinsson *et al.*, 2000; Chantal *et al.*, 2002; Christiansen *et al.*, 1995; Ernst *et al.*, 1997; Heun *et al.*, 1997; Parnetti *et al.*, 1997; Rose *et al.*, 1999; Schuff *et al.*, 1997; Watanabe *et al.*, 2002). Single voxel ^{1}H MRS studies of AD have consistently found reductions in NAA/Cr in the hippocampus/medial temporal lobe (Chantal *et al.*, 2002, 2004; Dixon *et al.*, 2002; Jessen *et al.*, 2000; Schuff *et al.*, 1997; Watanabe *et al.*, 2002) and other temporal lobe regions (Frederick *et al.*, 1997; Herminghaus *et al.*, 2003; Kantarci *et al.*, 2000; Parnetti *et al.*, 1997), and the midline parietal lobe/posterior cingulate (Antuono *et al.*, 2001; Griffith *et al.*, 2007a; Hattori *et al.*, 2002; Herminghaus *et al.*, 2003; Kantarci *et al.*, 2000, 2002b, 2003; Martinez-Bisbal *et al.*, 2004; Rose *et al.*, 1999), although two studies have failed to find abnormal NAA measures in this latter region (Waldman and Rai, 2003; Watanabe *et al.*, 2002). Findings of reduced NAA have also been seen in the temporal-parietal area (Chantal *et al.*, 2002; Ernst *et al.*, 1997; Parnetti *et al.*, 1996), occipital lobes (Kantarci *et al.*, 2000; Moats *et al.*, 1994; Shonk *et al.*, 1995; Waldman *et al.*, 2002; Watanabe *et al.*, 2002; Weiss *et al.*, 2003), and frontal lobes (Chantal *et al.*, 2002, 2004; Christiansen *et al.*, 1995; Herminghaus *et al.*, 2003; Parnetti *et al.*, 1997). In keeping with the emerging understanding of NAA as a marker for axonal viability as well as neuronal function (Valenzuela and Sachdev, 2001), some studies have found reduced NAA in parietal lobe white matter (Herminghaus *et al.*, 2003; Moats *et al.*, 1994) and subcortical white matter (Catani *et al.*, 2001; Hattori *et al.*, 2002; Heun *et al.*, 1997; Meyerhoff *et al.*, 1994), while others have not demonstrated this result (Catani *et al.*, 2002; Watanabe *et al.*, 2002). Studies using a whole brain spectral or multispectral approach have generally demonstrated reduced NAA in AD (Adalsteinsson *et al.*, 2000; Pfefferbaum *et al.*, 1999), which is mainly observed in posterior gray matter (Adalsteinsson *et al.*, 2000; MacKay *et al.*, 1996). Taken together, reports of widespread reductions in NAA are highly consistent with the known progression of neurofibrillary pathology of AD, which first is prevalent limbic cortical regions but then progresses to primary sensory-motor and visual cortices in more advanced stages (Braak and Braak, 1991).

A few studies to date have examined the pattern of changes in NAA over the course of one year in AD. Two such studies have pointed toward relative declines in NAA levels in AD patients (Adalsteinsson *et al.*, 2000; Kantarci *et al.*, 2007), while two others found no significant within-subjects declines (Dixon *et al.*, 2002; Jessen *et al.*, 2001) despite significantly lower NAA levels compared to older controls at both time points in one study (Dixon *et al.*, 2002). The indication of these few longitudinal studies is that NAA is potentially sensitive to the progression of AD (Doraiswamy *et al.*, 1998; Kantarci *et al.*, 2007), although more studies are desirable.

Some studies in AD patients have detected abnormal measurements of Cho (Chantal *et al.*, 2002, 2004; Jessen *et al.*, 2000; Kantarci *et al.*, 2000, 2003; Lazeyras *et al.*, 1998; MacKay *et al.*, 1996; Meyerhoff *et al.*, 1994), although results have been inconsistent, with some studies finding reduced Cho levels in AD patients compared to controls (Chantal *et al.*, 2002, 2004; Jessen *et al.*, 2000; Kantarci *et al.*, 2000, 2003) and other studies finding elevated Cho levels (Lazeyras *et al.*, 1998; MacKay *et al.*, 1996; Meyerhoff *et al.*, 1994). A longitudinal study found that that Cho was elevated in AD patients versus older controls at baseline, but that the annualized rate of change in Cho did not differ between these groups (Kantarci *et al.*, 2007). The reasons for differences in Cho findings across studies are unclear. Cho elevations may be a consequence of increased membrane turnover secondary to neurodegeneration. Alternatively, Cho levels may increase secondary to increased membrane phosphatidylcholine catabolism, which may serve as a compensatory mechanism against chronically reduced acetylcholine levels in AD (Kantarci *et al.*, 2007). Abnormal Cho levels in AD may also be due to reduced choline acetyltransferase activity secondary to cholinergic agonist treatment (Griffith *et al.*, 2008c; Kantarci *et al.*, 2007). Alternatively, discrepant findings across studies may be due to variations in ^{1}H MRS methods, with studies using ^{1}H MRS sequences with longer echo times tending to find elevated Cho levels (Lazeyras *et al.*, 1998; MacKay *et al.*, 1996), while those with shorter echo times finding lowered Cho levels (Chantal *et al.*, 2002, 2004; Kantarci *et al.*, 2000, 2003). The location of the voxel of interest is also likely to account for differences, with decreased Cho detected in the posterior cingulate (Kantarci *et al.*, 2000, 2003) and medial temporal lobe (Chantal *et al.*, 2002, 2004; Jessen *et al.*, 2000), while increased Cho was detected in the posterior gray matter (Lazeyras *et al.*, 1998; MacKay *et al.*, 1996) and gray matter of the centrum semiovale (Meyerhoff *et al.*, 1994).

Studies have frequently found abnormal elevations of mI in AD patients, most often in the temporal-parietal area (Chantal *et al.*, 2002, 2004; Ernst *et al.*, 1997; Parnetti *et al.*, 1996), posterior cingulate gyrus/mesial parietal lobe (Griffith *et al.*, 2007a; Herminghaus *et al.*, 2003; Kantarci *et al.*, 2000, 2002b, 2003; Lazeyras *et al.*, 1998; Martinez-Bisbal *et al.*, 2004; Rose *et al.*, 1999; Waldman and Rai, 2003), parietal white matter (Herminghaus *et al.*, 2003; Moats *et al.*, 1994), and occipital lobes (Moats *et al.*, 1994; Shonk *et al.*, 1995; Waldman *et al.*, 2002). Abnormal mI levels are detected less often in the frontal lobes (Chantal *et al.*, 2002, 2004; Herminghaus *et al.*, 2003; Parnetti *et al.*, 1997) and subcortical regions (Catani *et al.*, 2001, 2002; Hattori *et al.*, 2002; Heun *et al.*, 1997) in AD, which is in agreement with the known regional distribution of AD neuropathology. One study performed at 3 T failed to detect mI differences in the posterior cingulate in AD patients (Hattori *et al.*, 2002), possibly due to higher magnetic field inhomogeneity and susceptibility effects compared to 1.5 T (Kantarci *et al.*, 2003). A longitudinal study conducted at 1.5 T found no difference in annual rate of change in mI between AD patients versus older controls, although mI levels were elevated in AD patients at baseline (Kantarci *et al.*, 2007). Because of the robustness of findings involving decreased NAA and increased mI in AD patients, and because these two metabolites appear to represent independent markers of disease progression (Doraiswamy *et al.*, 1998), some researchers have used the ratio of NAA/mI to increase the sensitivity of ^{1}H MRS to metabolite changes in AD (Kantarci *et al.*, 2002b, 2003; Martinez-Bisbal *et al.*, 2004; Parnetti *et al.*, 1997; Rose *et al.*, 1999; Waldman and Rai, 2003; Weiss *et al.*, 2003). Reduced NAA/mI in AD versus healthy controls has been observed in the posterior cingulate (Kantarci *et al.*, 2002b, 2003), midline parietal lobe (Rose *et al.*, 1999), and frontal white matter (Parnetti *et al.*, 1997), but not frontal gray matter (Kizu *et al.*, 2004).

The majority of (Mueller *et al.*, 2006) ^{1}H MRS studies in AD have used Cr in the denominator of the metabolic ratio (Antuono *et al.*, 2001; Bartres-Faz *et al.*, 2002; Block *et al.*, 2002; Catani *et al.*, 2001, 2002; Chantal *et al.*, 2004; Christiansen *et al.*, 1995; Dixon *et al.*, 2002; Doraiswamy *et al.*, 1998; Ernst *et al.*, 1997; Frederick *et al.*, 1997; Hattori *et al.*, 2002; Herminghaus *et al.*, 2003; Heun *et al.*, 1997; Jessen *et al.*, 2000, 2001; Kantarci *et al.*, 2000, 2002b, 2003; Kattapong *et al.*, 1996; Lazeyras *et al.*, 1998; MacKay *et al.*, 1996; Martinez-Bisbal *et al.*, 2004; Meyerhoff *et al.*, 1994; Rose *et al.*, 1999; Schuff *et al.*, 1997; Shonk *et al.*, 1995; Waldman and Rai, 2003; Waldman *et al.*, 2002; Weiss *et al.*, 2003), due to its presumed invariance in brain disease (Valenzuela and Sachdev, 2001). Findings in AD have generally indicated no increases or decreases in Cr compared to controls (Ernst *et al.*, 1997; Pfefferbaum *et al.*, 1999; Schuff *et al.*, 1997). However, some have documented decreased Cr levels in AD patients versus normal controls in the left mesial temporal lobe, parieto-temporal cortex (Chantal *et al.*, 2002), and occipital lobe gray matter (Moats *et al.*, 1994).

The GLX peak has been investigated in a few studies of AD (Antuono *et al.*, 2001; Ernst *et al.*, 1997; Hattori *et al.*, 2002; Herminghaus *et al.*, 2003; Moats *et al.*, 1994), more recently with MRI scanners at field strengths other than 1.5 T (Antuono *et al.*, 2001; Hattori *et al.*, 2002). These studies have mostly reported reduced GLX levels in AD patients compared to controls in the posterior cingulate (Antuono *et al.*, 2001; Hattori *et al.*, 2002), occipital lobe (Moats *et al.*, 1994), and the dominant-hemisphere lateral temporal cortex (Herminghaus *et al.*, 2003), although two studies have failed to find GLX differences in patients with AD (Ernst *et al.*, 1997; Griffith *et al.*, 2008c).

Our group has performed one of the only studies to date systematically investigating sI in AD. We identified elevated sI/Cr in the posterior cingulate of AD patients, with MCI patients showing a trend toward elevation (Griffith *et al.*, 2007a); a prior study demonstrating raised concentrations of sI has been reported in the normal aging human brain (Kaiser *et al.*, 2005). The mechanism by which sI elevations occur in AD is unclear, although given that sI is a breakdown product of mI metabolism, this finding may reflect the same processes involved in increased mI. sI agonist treatment has been recently shown to inhibit AD-like brain pathology in mice (Fenili *et al.*, 2007).

Utility of (Mueller *et al.*, 2006) *^{1}H MRS for discriminating Alzheimer's disease from normal aging*: ^{1}H MRS has been used to discriminate AD patients from cognitively healthy older adults (Antuono *et al.*, 2001; Dixon *et al.*, 2002; Ernst *et al.*, 1997; Kantarci *et al.*, 2002b; MacKay *et al.*, 1996; Schuff *et al.*, 1997; Shonk *et al.*, 1995). Although an early study concluded that reproducibility and inter-rater error limited the utility of ^{1}H MRS as a clinical tool in AD (Heun *et al.*, 1997), several studies have since demonstrated that NAA can provide an adequate means of discriminating normal controls from AD patients (Antuono *et al.*, 2001; Schuff *et al.*, 1997; Shonk *et al.*, 1995) or improve discrimination of AD using structural neuroimaging variables (Dixon *et al.*, 2002; Ernst *et al.*, 1997; Kantarci *et al.*, 2002b; MacKay *et al.*, 1996). Among (Mueller *et al.*, 2006) ^{1}H MRS metabolite ratios, the NAA/mI ratio appears to best discriminate AD patients from normal controls (Kantarci *et al.*, 2007). The average accuracy of detecting AD with ^{1}H MRS is 87.75% (range 80-100%), and the average accuracy of identifying normal controls is 75.25 (range 73-78%) (Antuono *et al.*, 2001; Ernst *et al.*, 1997; MacKay *et al.*, 1996; Schuff *et al.*, 1997). Estimated sensitivity of ^{1}H MRS to AD is 83%, with a specificity of 95%, positive predictive value of 98% and a negative predictive value of 65% (Shonk *et al.*, 1995). Using a fixed specificity of 80%, one study reported a sensitivity of 82% for the NAA/mI posterior cingulate (Kantarci *et al.*, 2002b). An early longitudinal study that reported 70% correct classification of AD with ^{1}H MRS at baseline and 68% after 1 year for NAA of the left hippocampus (Dixon *et al.*, 2002). These results are consistent with a more recent longitudinal study examining the discriminatory value of NAA levels of the posterior cingulate in AD patients (Kantarci *et al.*, 2007).

^{1}H MRS and neurocognitive, psychological, and IADL measures in AD: Several studies have observed significant correlations between ^{1}H MRS and cognitive severity in AD as measured by a common mental status measure, the Mini Mental State Examination (MMSE) (Antuono *et al.*, 2001; Dixon *et al.*, 2002; Doraiswamy *et al.*, 1998; Ernst *et al.*, 1997; Heun *et al.*, 1997; Jessen *et al.*, 2000, 2001; Parnetti *et al.*, 1997; Rose *et al.*, 1999; Waldman and Rai, 2003); although, other studies have found no association between ^{1}H MRS and MMSE (Block *et al.*, 2002; Hattori *et al.*, 2002; Watanabe *et al.*, 2002). Fewer studies to date have investigated the relationships between other neurocognitive measures and ^{1}H MRS in AD. Pfefferbaum and colleagues (Pfefferbaum *et al.*, 1999) reported negative correlations of Cr with word recognition memory and negative correlations of Cho with face recognition memory. Chantal and colleagues (Chantal *et al.*, 2002) observed significant positive correlations between NAA of the left medial temporal lobe and verbal learning and recall. Negative correlations were detected between left

parieto-temporal cortex Cr and a confrontational naming measure, and right parieto-temporal cortex mI and a visual construction measure (Chantal *et al.*, 2002). One longitudinal study observed that in AD patients the annual percent change in NAA/Cr ratios was positively associated with annual change on a commonly used mental status measure, the Dementia Rating Scale (DRS), and negatively associated with change in a dementia severity rating scale, the Clinical Dementia Rating (CDR) (Kantarci *et al.*, 2007). These correlations were of comparable magnitude with correlations observed between the rate of change in MRI ventricular expansion and clinical progression measures (Kantarci *et al.*, 2007).

Other studies have investigated relationships between IADLs and ^{1}H MRS measures in AD. In a sample of mild AD patients, our group has demonstrated that NAA/Cr shows a positive correlation with a measure of financial abilities, while Cho/Cr showed a negative correlation with this same financial measure (Griffith *et al.*, 2007c). Antuono and colleagues (Antuono *et al.*, 2001) observed a significant correlation between GLX ratios from the posterior cingulate and an informant-report measure of daily activities in patients with AD, indicating more functional impairment as the GLX ratios decreased.

Postmortem neuropathologic correlates of in vivo ^{1}H MRS in AD: A recent study examined postmortem neuropathological correlates of antemortem ^{1}H MRS of the posterior cingulate and inferior precuneate gyri in healthy controls and in low- or high-likelihood AD, who were classified based on standard neuropathologic AD criteria (Kantarci *et al.*, 2008). The NAA/Cr, mI/Cr, and NAA/mI ratios differed between the low-, intermediate-, and high-likelihood AD groups after adjusting for age, sex, and time from ^{1}H MRS to death. Moreover, NAA/mI showed the strongest association with Braak neurofibrillary pathology staging, suggesting that NAA and mI provide complimentary information regarding the extent of disease involvement in AD patients. These results are consistent with a prior study reporting a correlation between postmortem ^{1}H MRS brain metabolite measures and neurofibrillary tangles and senile plaques in brain tissue of AD patients (Klunk *et al.*, 1998). Thus, antemortem ^{1}H MRS in patients with probable AD appears to be a sensitive marker of underlying AD neuropathology.

B. ^{1}H MRS Findings in Amnestic Mild Cognitive Impairment

Amnestic MCI has emerged as the diagnostic classification to denote the transitional phase between normal cognitive aging and AD (Petersen *et al.*, 2001). Diagnostic criteria for amnestic MCI include (1) subjective complaints of memory loss, preferably confirmed by an informant, (2) objective impairment on memory testing compared to age and educationally matched normative data, (3) otherwise normal performance on other cognitive tests, and (4) generally preserved activities of daily living (Petersen *et al.*, 2001). Individuals with MCI progress to clinically probable AD more frequently than cognitively normal peers (Ganguli *et al.*, 2004; Petersen *et al.*, 2001), with a rate of 10-17% annual progression compared to a rate of 1-2% per year in the general older adult population (Ganguli *et al.*, 2004; Griffith *et al.*, 2006a; Petersen *et al.*, 1999, 2000, 2001; Storandt *et al.*, 2002; Tierney *et al.*, 1996).

Several cross-sectional (Mueller *et al.*, 2006) ^{1}H MRS studies have been performed to date in patients with amnestic MCI. Kantarci and colleagues (Kantarci *et al.*, 2000) examined metabolic ratios in the left temporal lobe, the posterior cingulate, and the medial occipital lobe of MCI and AD patients. Patients with MCI differed from controls only in significant increases of mI/Cr in the posterior cingulate; compared to AD patients, MCI patients had significantly lower cingulate mI/Cr ratios. The authors speculated that increased mI in MCI patients reflected very early pathological changes (possibly gliosis) preceding neuronal dysfunction, as would be indicated by decrements in NAA. Subsequent ^{1}H MRS studies in MCI patients have confirmed mI increases compared to controls in the posterior cingulate (Kantarci *et al.*, 2003; Rami *et al.*, 2007a), left hippocampus (Franczak *et al.*, 2007), and other brain regions such as the bilateral paratrigonal white matter (Catani *et al.*, 2001) and parieto-temporal cortex (Chantal *et al.*, 2004; Rami *et al.*, 2007a). Other studies, however, have not found differences between mI levels in AD and MCI (Catani *et al.*, 2001; Chantal *et al.*, 2004; Garcia Santos *et al.*, 2008; Kantarci *et al.*, 2003).

Other (Mueller *et al.*, 2006) ^{1}H MRS metabolic abnormalities in MCI patients include a significant increase of Cho in the right frontal cortex and posterior cingulate (Chantal *et al.*, 2004; Kantarci *et al.*, 2003), in addition to Cho and NAA decreases in the left medial temporal lobe (Chantal *et al.*, 2004). NAA decreases were also found in the right hippocampus in a small group of MCI patients (Franczak *et al.*, 2007) and were reduced in the posterior cingulate of MCI patients compared to controls, but not as severe as seen in AD patients (Kantarci *et al.*, 2003). In contrast, Ackl *et al.* (2005) detected equally reduced NAA/Cr in the hippocampus of MCI patients and AD patients, although posterior cingulate metabolic abnormalities were observed only in the AD patients. Other studies have not detected any NAA changes in MCI (Garcia Santos *et al.*, 2008; Kantarci *et al.*, 2002a). Thus, the cross-sectional data generally indicates that metabolic brain abnormalities in MCI are transitional between normal metabolism observed in healthy older adults and abnormalities seen in AD. The pattern of findings generally concurs with the known early neuropathology of AD (Braak and Braak, 1991; Markesbery *et al.*, 2006).

To our knowledge, only two studies have investigated longitudinal metabolic changes measured by ^{1}H MRS in MCI patients (Bartnik Olson *et al.*, 2008; Kantarci *et al.*, 2007). Kantarci and colleagues (Kantarci *et al.*, 2007) reported declines in NAA/Cr of the posterior cingulate in MCI and AD patients on scans occurring approximately 1 year apart. There was a similar rate of decline in NAA for MCI patients who converted to AD versus those patients who remained stable.

No changes were observed in mI/Cr. Interestingly, Cho/Cr levels declined in MCI patients who remained stable, suggesting to these authors that a compensatory cholinergic mechanism explained the stability of these non-converter MCI patients. This result appears consistent with the recent finding that MCI converters exhibited elevated Cho/Cr compared to non-converters at baseline (Rami *et al.*, 2007b). In a second longitudinal study, Bartnik Olson and colleagues (Bartnik Olson *et al.*, 2008) found increases of mI concentrations in the posterior cingulate gyrus occurring over an average interval of 11 months, but non-significant interval increases in NAA or Cho concentrations. Discrepant findings between these two longitudinal studies may be due to the fact that Bartnik *et al.* used a method to model spectral components to quantify metabolite levels, while Kantarci and colleagues used metabolic ratios as obtained automatically following acquisition on a General Electric (GE) scanner.

^{1}H MRS and neurocognitive measures in MCI: Kantarci and colleagues (Kantarci *et al.*, 2002a) performed correlations between a global measure of cognition (DRS) and a measure of auditory verbal learning (AVLT) with posterior cingulate ^{1}H MRS ratios of NAA/Cr, NAA/mI, and mI/Cr. All three ratios were significantly correlated with the DRS and to a lesser extent with the AVLT. Ackl and colleagues indicated that hippocampal NAA/Cr correlated with verbal fluency and confrontation naming, but interestingly not with memory, in a mixed MCI and AD group (Ackl *et al.*, 2005). Posterior cingulate ^{1}H MRS correlated with a wider range of cognitive measures. We very recently reported that visual executive function, as measured by clock drawing, was negatively correlated with mI/Cr in amnestic MCI patients (Griffith *et al.*, 2007b). Furthermore, mI/Cr ratios accounted for 26% of the variance in clock drawing scores after accounting for overall cognition, age, and other potential confounding variables through stepwise multiple regression analysis.

^{1}H MRS predictors of dementia in MCI: Studies have indicated that baseline NAA/Cr can predict subsequent progression, or "conversion," from MCI to AD with high sensitivity and specificity. A recent study indicated that baseline NAA/Cr of the occipital lobe was able to discriminate amnestic MCI converters from non-converters over a mean follow-up period of 3 years, with 100% sensitivity and 75% specificity (Modrego *et al.*, 2005). Another recent study reported that MCI patients who converted to dementia within a 1-year follow-up showed reduced NAA/Cr of the left paratrigonal white matter compared to stable MCI patients (Metastasio *et al.*, 2006). A third study concluded that NAA/Cr ratios in both the posterior cingulate gyrus and left occipital cortex predicted conversion with a sensitivity of about 80% and a specificity of about 70%. In general, these studies suggest that ^{1}H MRS may prove valuable in identifying patients with MCI who are at risk for dementia conversion, although participant selection issues limit the validity of the current studies, including the operational criteria for MCI (Modrego *et al.*, 2005) and extremely high conversion rates over 1 year (Metastasio *et al.*, 2006).

C. ^{1}H MRS Findings in Frontotemporal Dementia

FTD refers to a heterogeneous group of clinical syndromes involving progressive neurodegeneration of the frontal and/or temporal lobes, including gliosis, neuron loss, and vacuolation of surface cortical layers (Coulthard *et al.*, 2006; Ernst *et al.*, 1997).

Findings of metabolic profiles in FTD patients compared to controls have shown metabolic abnormalities in many brain cortical regions. One study investigated (Mueller *et al.*, 2006) ^{1}H MRS located in the temporal, frontal, and parietal lobes and found lower ratios of NAA/Cr in the temporal and frontal lobe voxels, in addition to an increased mI/Cr ratio in the frontal lobe voxel (Coulthard *et al.*, 2006). No ^{1}H MRS abnormalities were observed in the parietal lobes (Coulthard *et al.*, 2006). Other ^{1}H MRS studies suggest that the spatial distribution of brain metabolic abnormalities of FTD patients differs from that of AD patients. One such study found reduced NAA and Glx and increased mI in mid-frontal gray matter in FTD patients versus AD patients and healthy controls, whereas a trend toward significantly increased mI in the temporoparietal gray matter was observed in AD versus FTD patients (Ernst *et al.*, 1997; also see Mihara *et al.*, 2006).

Studies have reported the correspondence of ^{1}H MRS abnormalities with clinical and cognitive measures in patients with FTD, highlighting the clinical importance of these metabolic abnormalities. Clinical/functional measures (i.e., CDR and the MMSE) have been associated with NAA and mI in the frontal region and mI and Glx in the temporopartietal region in a group of FTD and AD patients and healthy controls (Ernst *et al.*, 1997). Rami *et al.* (2008) also reported associations of naming ability with NAA/Cr and Cho/Cr ratios in the left temporal pole, but not the left temporoparietal region, after controlling for MMSE scores across a sample of FTD, MCI, and AD patients, and heathy controls.

D. ^{1}H MRS Findings in Vascular Dementia

VAD results from one or a combination of cerebrovascular mechanisms that cause degenerative changes in the brain, including infarcts, white matter lesions, and resultant atrophy (Wiederkehr *et al.*, 2008). Diagnosis of VAD presents a challenge clinically due to its etiological heterogeneity, variable clinical manifestation, and co-occurrence with AD pathology (only 10% of dementia cases show pure vascular pathology on postmortem brain examination) (Holmes *et al.*, 1999; Jones and Waldman, 2004).

^{1}H MRS has revealed widespread brain metabolic abnormalities in VAD patients. In one study, Herminghaus *et al.* (2003) measured spectra from voxels in five brain regions: mid-parietal gray and white matter, mid-frontal gray and white matter, and the temporal gyrus mixed gray/white matter. Compared to healthy controls, VAD patients exhibited global reductions in NAA/Cr ratios in all voxels, along with elevations in mI/Cr ratios in parietal gray and white matter, frontal white matter, and the temporal lobe, and elevations in Glx ratios in parietal gray matter and the temporal lobe. In another study, Kantarci and colleagues (Kantarci *et al.*, 2004) reported that in the posterior cingulate VAD patients had lower ratios of NAA/Cr, but normal ratios of mI/Cr and Cho/Cr, when compared with healthy controls. These studies suggest that VAD involves multiple neuropathological processes, even in regions distant from actual vascular pathology (Kantarci *et al.*, 2004), including neuronal dysfunction/loss, axonal damage, myelin loss, and gliosis (Jones and Waldman, 2004). Kantarci *et al.* (2004) have speculated that reductions in NAA in distal brain areas may be secondary to retrograde Wallerian degeneration caused from brain regions showing actual vascular neuropathology.

E. ¹H MRS Findings in Dementia with Lewy Bodies

The cardinal features of DLB include progressive cognitive decline, along with fluctuations in alertness and attention, persistent visual hallucinations, and spontaneous parkinsonian motor symptoms, including rigidity and the loss of spontaneous movement (McKeith *et al.*, 2005). Of note, DLB patients exhibit a striking reduction in cortical acetylcholine activity relative to AD patients (Tiraboschi *et al.*, 2000).

A relative paucity of ¹H MRS studies has examined patients with DLB, and their findings have been variable. Molina *et al.* (2002) initial study in this patient population found decreased NAA/Cr, Cho/Cr, and Glx/Cr ratios in DLB patients versus healthy controls in white matter from the left centrum semiovale, but did not find significant metabolite differences in gray matter from the midline parietal region. This study also found no correlations of any metabolite levels with clinical measures (age at onset and disease duration) or measures of cognitive and motor impairment, a result that is consistent with a prior neurochemical study which failed to find a relationship between reductions in acetylcholine activity and cognitive impairment in DLB (Tiraboschi *et al.*, 2000). In contrast, a more recent study found elevated NAA/Cr ratios in the left and right hippocampus in DLB patients versus controls but did not find group differences in Cho/Cr ratios, although their sample size was somewhat limited ($n = 8$ in both groups) (Xuan *et al.*, 2008). Kantarci *et al.* (2004) reported elevated Cho/Cr ratios in a voxel encompassing the posterior cingulate gyrus of DLB patients versus healthy controls, but did not find group differences in NAA/Cr or mI/Cr ratios.

The data on ¹H MRS findings in DLB are sparse and variable. In some respects, these variable findings are similar to those of postmortem pathological studies on DLB, which have yielded inconsistent findings regard to neuronal loss in DLB (Cordato *et al.*, 2000; Gomez-Isla *et al.*, 1999). Moreover, methodological differences, including differences in symptom severity of patients, voxel placement, and (Mueller *et al.*, 2006) ^{1}H MRS acquisition parameters (e.g., short versus long echo time), may also account for some of the variable results across these studies.

F. ^{1}H MRS Findings in Parkinson's Disease Dementia

Dementia observed in patients with Parkinson's disease can manifest from a variety of clinical and neuropathological entities, including diffuse Lewy body pathology (Galvin *et al.*, 2006), AD (Galvin *et al.*, 2006), and isolated Lewy body pathology of the substantia nigra (Galvin *et al.*, 2006; Pletnikova *et al.*, 2005). A specific dementia in patients with Parkinson's disease, termed PD with dementia (PDD), involves the onset of a progressive dementing illness at least 1 year subsequent to the onset of the movement disorder (McKeith *et al.*, 2005). PDD is believed to be clinically (Benecke, 2003), cognitively (Aarsland *et al.*, 2003), neuropsychologically (Litvan *et al.*, 1991), and neuropathologically (Kovari *et al.*, 2003) distinct from that of patients with DLB (Benecke, 2003) and patients with AD and late motor complications (Dickson, 2000).

^{1}H MRS findings indicate that PDD patients exhibit abnormal brain metabolism as compared against healthy controls and nondemented PD patients. Initial ^{1}H MRS studies in PDD examined metabolite levels in the occipital cortex. These studies found reductions in NAA levels in PDD patients versus nondemented PD patients, but not healthy controls (Summerfield *et al.*, 2002), and elevations in lactate levels in PDD patients versus both nondemented PD patients and healthy controls (Bowen *et al.*, 1995), although similar lactate abnormalities have not been documented subsequently (Firbank *et al.*, 2002). More recent ^{1}H MRS studies by our group have investigated metabolism of the posterior cingulate gyrus, which shows neuropathological changes in postmortem PDD studies (Braak *et al.*, 2004) and abnormal blood flow (Osaki *et al.*, 2005) and dopamine transmitter alterations (Brooks and Piccini, 2006) in antemortem PDD imaging studies. Consistent with these post- and antemortem studies, ^{1}H MRS of the posterior cingulate gyrus in PDD has revealed reduced NAA/Cr ratios in PDD patients versus healthy controls (Griffith *et al.*, 2008a,c) and nondemented PD patients (Griffith *et al.*, 2008a), in addition to reduced Glu/Cr versus healthy controls (Griffith *et al.*, 2008c). Whereas lower NAA levels presumably reflect compromised functional integrity of neurons in the posterior cingulate gyrus and occipital lobe in PDD, lower Glu reductions could plausibly reflect other aspects of the PDD disease process, such as downregulation of glutaminergic cortical efferents from the basal ganglia to the posterior cingulate gyrus (Griffith *et al.*, 2008c).

Of note, ^{1}H MRS-derived NAA and Glu measures in the posterior cingulate gyrus appear to be able to discriminate PDD patients from nondemented PD patients. The NAA/Cr ratio has shown high sensitivity and reasonable specificity in correctly discriminating individuals with PDD from PD and healthy controls (Griffith *et al.*, 2008a). Moreover, both NAA and Glu levels have shown modestly strong correlations with mental status measures (MMSE and DRS scores) in PDD (Griffith *et al.*, 2008a,c). Future longitudinal ^{1}H MRS studies determining whether these brain metabolic abnormalities precede the clinical manifestation of PDD appear warranted.

G. Utility of ^{1}H MRS for Discriminating Among Dementias

Studies have sought to compare the sensitivity and specificity of brain metabolic profiles in discriminating among neurodegenerative dementias. Most all of these studies have compared patients with AD to another dementia, a question that is of clinical value given the high prevalence of AD (McMurtray *et al.*, 2006). Ernst and colleagues (Ernst *et al.*, 1997) found that FTD patients could be discriminated from AD patients and controls using a combination of NAA and Cr values from the mid-frontal and left temporoparietal region, with accuracy of detecting FTD of 92% and an accuracy of detecting AD of 82%. Shonk *et al.* (1995) indicated that NAA/Cr of the occipital lobe detected AD from FTD with a sensitivity of 82%, specificity of 64%, positive predictive value of 74% and a negative predictive value of 80%. Other studies have found no differences on (Mueller *et al.*, 2006) ^{1}H MRS between AD and FTD patients. For example, a recent study found reduced NAA/Cr levels and increased mI/Cr levels in the posterior cingulate gyrus in both FTD patients and AD patients when compared with healthy controls (Mihara *et al.*, 2006). Several other studies have found that metabolite concentrations did not differ significantly between FTD and AD patients (Garrard *et al.*, 2006; Kantarci *et al.*, 2004; Kizu *et al.*, 2004).

Other studies have attempted to distinguish between VAD and AD using ^{1}H MRS. Patients with VAD, when compared to patients with AD, show reduced NAA/Cr ratios in frontal gray matter and subcortical white matter, and elevated mI/Cr in frontal white matter (Herminghaus *et al.*, 2003; Kattapong *et al.*, 1996; Martinez-Bisbal *et al.*, 2004). In comparison, patients with AD show increased mI/Cr in the posterior cingulate gyrus (Kantarci *et al.*, 2004; Martinez-Bisbal *et al.*, 2004) and occipital lobe (Rai *et al.*, 1999; Waldman *et al.*, 2002), as well as increased Cho/Cr of the posterior cingulate (Martinez-Bisbal *et al.*, 2004), when compared with VAD patients. One study reported an 83% classification accuracy for AD and 71% accuracy for VaD using NAA measures from whole brain gray matter (MacKay *et al.*, 1996), while a second study comparing patients with

AD, VaD, MCI, and major depression reported that the NAA/mI ratio had an 81.5% sensitivity and 72.7% specificity for discriminating AD from the other groups (Martinez-Bisbal *et al.*, 2004).

We are aware of only two studies to date that have compared ^{1}H MRS across patients with Parkinson-spectrum dementias to patients with other dementias. Kantarci and colleagues (Kantarci *et al.*, 2004) reported that patients with DLB had normal NAA/Cr ratios when compared to patients with AD and patients with FTD; patients with DLB also had normal mI/Cr ratios compared to patients with FTD. No differences were observed between DLB patients and patients with VAD. Griffith and colleagues (Griffith *et al.*, 2008c) very recently reported that patients with PDD showed abnormally low Glu/Cr ratios when compared with patients with AD, while patients with AD showed a trend toward higher mI/Cr ratios when compared with patients with PDD. The Glu/Cr ratio was able to discriminate between AD and PDD at rates greater than chance. No studies to date that we are aware of have compared (Mueller *et al.*, 2006) ^{1}H MRS between patients with DLB versus patients with PDD, although both dementias show considerable clinical and neuropathological overlap (Galvin *et al.*, 2006).

H. Utility of ^{1}H MRS for Monitoring Treatment Effects in Dementias

A few studies have used ^{1}H MRS to assess responses to pharmacological treatment in AD patients. Krishnan *et al.* (2003) reported that NAA levels in subcortical gray matter and periventricular white matter increased following 12 weeks of therapy with donepezil, a cholinesterase inhibitor. Jessen and colleagues (Jessen *et al.*, 2006) observed changes in NAA/Cr, Cho/Cr and mI/Cr ratios associated with donepezil treatment; the ^{1}H MRS changes were also associated with improvement on the Alzheimer's Disease Assessment Scale-cognitive subscale (ADAS-cog), a commonly used measure of treatment effects in AD. In contrast, Bartha *et al.* (2008) found a decrease in NAA levels, along with a decrease in mI levels, in the right hippocampus following 16 weeks of donepezil treatment. In another study, 4 months of treatment with the cholinesterase inhibitor rivastigmine was associated with an increase in NAA/Cr in the frontal lobe in AD patients versus untreated AD patients; there was an association reported between the metabolite changes and modest clinical improvement in the treatment group (Modrego *et al.*, 2006). A study investigating the effects of treatment with xanomeline, a cholinergic agonist, have found associations between pre- and posttreatment changes in NAA/Cr, Cho/Cr, and mI/Cr ratios and changes in ADAS-cog scores (Frederick *et al.*, 2002). These studies suggest that ^{1}H MRS has the potential to monitor disease modifying treatment effects that are correlated with clinical outcome measures.

The ability of ^{1}H MRS to quantify therapeutic effects in dementias other than AD is unknown. One study investigated the effect of argatroban therapy, a potent selective thrombin inhibitor, on brain metabolite levels in peripheral arterial occlusive disease patients, two of whom were diagnosed with VAD (Kario *et al.*, 1999). Prior to argatroban therapy, patients with silent cerebrovascular disease had decreased NAA/Cr in the deep white matter. However, NAA/Cr ratios increased significantly in patients with silent cerebrovascular disease following argatroban therapy, accompanied by improved scores on the MMSE and improvements in activities of daily living in the two VAD patients. While this study highlights the potential importance of ^{1}H MRS in the monitoring of therapeutic effects, continued study in larger samples of VAD and with other therapeutic agents is clearly needed.

IV. Discussion

The use of ^{1}H MRS as a clinical investigative tool in neurodegenerative dementias has yielded a body of findings that could prove important for research into the causes and treatments of dementias. While there have been some methodological and technical advances in ^{1}H MRS over the years, there have been relatively consistent findings in studies of AD, providing evidence of the stability of this technique as a potential biomarker. The emergence of studies in other dementias, and comparative studies across dementias, has also proved worthwhile. Nonetheless, several important issues remain to be addressed.

Across ^{1}H MRS studies, some common findings have emerged that engender confidence in the interpretation of brain metabolic abnormalities in dementias. One such finding has been the nearly ubiquitous observation of reduced cortical NAA in almost all of the major dementing illnesses (with the notable exception of DLB in one study (Kantarci *et al.*, 2004)). Reduction in NAA has been observed in all major lobes of the brain as well as in the hippocampus. While NAA reduction is primarily observed in gray matter, white matter NAA abnormalities have also been reported. These results reify the interpretation of NAA as a marker of neuronal and axonal integrity in dementias. NAA ratios have also shown strong correspondence with postmortem neuropathology in AD (Kantarci *et al.*, 2008), presumably because higher levels of AD neuropathology correspond to greater loss of neuronal function. However, as reduced NAA is observed in dementias such as FTD, where there is no AD neuropathology, it is clear that NAA abnormalities are not specific effects of AD neuropathology alone, but are rather sensitive to any form of neuropathology that is detrimental to neuronal functioning. The use of NAA in combination with other ^{1}H MRS metabolites (such as mI) might improve the specificity of ^{1}H MRS to detect specific forms of neuropathology (Kantarci *et al.*, 2008). An advantage of ^{1}H MRS is that data can be collected on several metabolites simultaneously, and thus can be used to determine metabolic profiles that could discriminate among dementias (Griffith *et al.*, 2008c; Kantarci *et al.*, 2004).

The sensitivity of ^{1}H MRS to detect dementias has been investigated in a few studies, showing generally good ability to discriminate between dementias and normally functioning individuals. However, ^{1}H MRS may hold the most promise as a means to identify earlier (preclinical) patients at risk for dementia, or completely asymptomatic individuals, who could then be the target for early intervention studies. The emerging body of findings in patients with amnestic MCI, some of whom progress to have clinically probable AD, indicates that ^{1}H MRS might be sensitive to the metabolic effects of neurodegenerative diseases prior to their clinical identification. An exciting area of development has been the identification of ^{1}H MRS abnormalities in patients who are at genetic risk of AD but who are otherwise completely asymptomatic, suggesting that brain metabolic abnormalities predate identification of cognitive deficits in these patients (Godbolt *et al.*, 2006). Cortical ^{1}H MRS changes that occur in patients with PD and no dementia (Griffith *et al.*, 2008b) resemble those reported in PDD patients (Griffith *et al.*, 2008c), suggesting that ^{1}H MRS may be helpful in determining risk of dementia in PD.

Longitudinal studies are slowly emerging to support the use of ^{1}H MRS as a means to track progressive brain metabolic changes. At least one study has demonstrated that ^{1}H MRS can be reliably obtained within the same participants at different time points (Rose *et al.*, 1999). A few longitudinal studies have been conducted to date in AD and MCI, although other dementias have not yet been investigated with longitudinal ^{1}H MRS. The findings of these longitudinal studies suggest that change in NAA corresponds well with change in clinical status of AD, and thus possibly could be used to measure disease progression with less error than is seen with clinical instruments such as the ADAS-cog (Mueller *et al.*, 2006). Some studies have employed ^{1}H MRS to measure treatment effectiveness in AD and shown links to modest cognitive improvement with improving metabolism. However, correspondence with clinical outcomes other than cognitive change, such as status of daily functional activities, is desirable to anchor ^{1}H MRS changes to real-world outcomes. Two cross-sectional studies have reported a correspondence between ^{1}H MRS and daily activities in AD (Antuono *et al.*, 2001; Griffith *et al.*, 2007c); these findings suggest that change in ^{1}H MRS would correspond to declines in complex daily activities. Clearly, a better understanding of the longitudinal course of brain metabolic abnormalities in dementias, and the associations of these metabolic abnormalities with cognitive and daily functional changes, is essential to pursuing ^{1}H MRS as a measure of treatment outcomes.

The search for viable biomarkers of dementing illnesses is critical for development of new treatments. Mueller and colleagues outlined their six criteria for a viable biomarker of treatment effects in patients with dementias: (1) links between the markers and treatment outcomes, (2) test-retest reliability; (3) sensitivity to desired stage of treatment; (4) link to the biomarker and amyloid burden, (5) non-invasiveness and tolerability; and (6) availability and cost (Mueller *et al.*, 2006). In considering ^{1}H MRS in light of these criteria, the current state of the literature suggests that several of the criteria are being approached. ^{1}H MRS appears to have some links to potential measurable outcomes, in that studies have reported associations with cognitive functioning and instrumental daily activities, as well as showing sensitivity to treatment effects. Test-retest reliability has been supported in at least one study, although interscan variability may not be at the desired level for reliable use in individual patients. Studies in ^{1}H MRS of preclinical dementias have shown that early metabolic changes are observable at the stage when treatment is most desirable (Petersen, 2004). ^{1}H MRS is arguably among one of the most widespread imaging modalities, in that many of the MRI scanners commercially available have the capability of collecting ^{1}H MRS data, although subsequent data processing and interpretation require expertise. In our experience, ^{1}H MRS is well tolerated in patients with mild to moderate stages of dementia as well as individuals with movement disorders.

The link of ^{1}H MRS with amyloid burden, one of the Mueller *et al.* (2006) criteria, is questionable. While amyloid deposition is clearly correlated with NAA and mI levels in patients with AD, abnormalities in these metabolites are also seen in non-amyloid neuropathologies such as FTD. ^{1}H MRS can likely play a role as a marker of neuronal integrity with sensitivity to a number of neuropathologies (i.e., Lewy bodies, tauopathies, cerebrovascular disease) rather than having a specificity to amyloid. Given that questions have arisen as to the key role that amyolid deposition plays in the dementing process in AD (Arends *et al.*, 2000; Dickson, 1997; Terry *et al.*, 1991), the importance of an amyloid biomarker over a biomarker of neuronal health is debatable.

Future directions in ^{1}H MRS of dementias should focus on developing stable and reliable longitudinal techniques and applying these to investigate longitudinal changes, especially in dementias other than AD as well as longitudinal comparative studies of AD with other dementias. Measuring correspondence of change on clinical and functional measures with changes in ^{1}H MRS measures is also needed. In addition, studies are needed to further investigate whether ^{1}H MRS measures are sensitive to medication effects, such as cholinesterase inhibitor use, in patients where these treatments are considered to be beneficial. The association of ^{1}H MRS measures with imaging of amyloid, such as comparison with the Pittsburgh B Compound PET procedure (Klunk *et al.*, 2004), would be beneficial. Lastly, as technological advances occur it would be worthwhile to expand from single-voxel approaches to 2D and 3D ^{1}H MRS imaging to better understand the extent of ^{1}H MRS abnormalities in dementias. Such developments will hopefully prove to benefit the search for effective treatments of dementias in the twenty-first century.

Acknowledgments

Funding was provided by grants from the National Institute of Aging (Alzheimer's Disease Research Center—1P50 AG16582-01: Harrell, PI; 1R01 AG021927-01: Marson, PI) and Alzheimer's of Central Alabama.

References

Aarsland, D., Litvan, I., Salmon, D., Galasko, D., Wentzel-Larsen, T., and Larsen, J. P. (2003). Performance on the dementia rating scale in Parkinson's disease with dementia and dementia with Lewy bodies: Comparison with progressive supranuclear palsy and Alzheimer's disease. *J. Neurol. Neurosurg. Psychiatry* **74**(9), 1215–1220.

Ackl, N., Ising, M., Schreiber, Y. A., Atiya, M., Sonntag, A., and Auer, D. P. (2005). Hippocampal metabolic abnormalities in mild cognitive impairment and Alzheimer's disease. *Neurosci. Lett.* **384**(1–2), 23–28.

Adalsteinsson, E., Sullivan, E. V., Kleinhans, N., Spielman, D. M., and Pfefferbaum, A. (2000). Longitudinal decline of the neuronal marker N-acetyl aspartate in Alzheimer's disease. *Lancet* **355**(9216), 1696–1697.

Antuono, P. G., Jones, J. L., Wang, Y., and Li, S. J. (2001). Decreased glutamate + glutamine in Alzheimer's disease detected *in vivo* with ^{1}H-MRS at 0.5 T. *Neurology* **56**(6), 737–742.

Arends, Y. M., Duyckaerts, C., Rozemuller, J. M., Eikelenboom, P., and Hauw, J. J. (2000). Microglia, amyloid and dementia in alzheimer disease. A correlative study. *Neurobiol. Aging* **21**(1), 39–47.

Bartha, R., Smith, M., Rupsingh, R., Rylett, J., Wells, J. L., and Borrie, M. J. (2008). High field (1)H MRS of the hippocampus after donepezil treatment in Alzheimer disease. *Prog. Neuropsychopharmacol. Biol. Psychiatry* **32**(3), 786–793.

Bartnik Olson, B. L., Holshouser, B. A., Britt, W., 3rd, Mueller, C., Baqai, W., Patra, S., Petersen, F., and Kirsch, W. M. (2008). Longitudinal Metabolic and Cognitive Changes in Mild Cognitive Impairment Patients. *Alzheimer. Dis. Assoc. Disord.*

Bartres-Faz, D., Junque, C., Clemente, I. C., Lopez-Alomar, A., Bargallo, N., Mercader, J. M., and Moral, P. (2002). Relationship among (1)H-magnetic resonance spectroscopy, brain volumetry and genetic polymorphisms in humans with memory impairment. *Neurosci. Lett.* **327**(3), 177–180.

Benecke, R. (2003). Diffuse Lewy body disease – a clinical syndrome or a disease entity? *J. Neurol.* **250** (Suppl 1), 39–42.

Block, W., Jessen, F., Traber, F., Flacke, S., Manka, C., Lamerichs, R., Keller, E., Heun, R., and Schild, H. (2002). Regional N-acetylaspartate reduction in the hippocampus detected with fast proton magnetic resonance spectroscopic imaging in patients with Alzheimer disease. *Arch. Neurol.* **59**(5), 828–834.

Bowen, B. C., Block, R. E., Sanchez-Ramos, J., Pattany, P. M., Lampman, D. A., Murdoch, J. B., and Quencer, R. M. (1995). Proton MR spectroscopy of the brain in 14 patients with Parkinson disease. *AJNR Am. J. Neuroradiol.* **16**(1), 61–68.

Braak, H., and Braak, E. (1991). Neuropathological stageing of Alzheimer-related changes. *Acta Neuropathol. (Berl)* **82**(4), 239–259.

Braak, H., Ghebremedhin, E., Rub, U., Bratzke, H., and Del Tredici, K. (2004). Stages in the development of Parkinson's disease-related pathology. *Cell Tissue Res.* **318**(1), 121–134.

Brand, A., Richter-Landsberg, C., and Leibfritz, D. (1993). Multinuclear NMR studies on the energy metabolism of glial and neuronal cells. *Dev. Neurosci.* **15**(3–5), 289–298.

Brooks, D. J., and Piccini, P. (2006). Imaging in Parkinson's disease: The role of monoamines in behavior. *Biol. Psychiatry* **59**(10), 908–918.

Catani, M., Cherubini, A., Howard, R., Tarducci, R., Pelliccioli, G. P., Piccirilli, M., Gobbi, G., Senin, U., and Mecocci, P. (2001). (1)H-MR spectroscopy differentiates mild cognitive impairment from normal brain aging. *Neuroreport* **12**(11), 2315–2317.

Catani, M., Mecocci, P., Tarducci, R., Howard, R., Pelliccioli, G. P., Mariani, E., Metastasio, A., Benedetti, C., Senin, U., and Cherubini, A. (2002). Proton magnetic resonance spectroscopy reveals similar white matter biochemical changes in patients with chronic hypertension and early Alzheimer's disease. *J. Am. Geriatr. Soc.* **50**(10), 1707–1710.

Chantal, S., Braun, C. M., Bouchard, R. W., Labelle, M., and Boulanger, Y. (2004). Similar ^{1}H magnetic resonance spectroscopic metabolic pattern in the medial temporal lobes of patients with mild cognitive impairment and Alzheimer disease. *Brain Res.* **1003**(1–2), 26–35.

Chantal, S., Labelle, M., Bouchard, R. W., Braun, C. M., and Boulanger, Y. (2002). Correlation of regional proton magnetic resonance spectroscopic metabolic changes with cognitive deficits in mild Alzheimer disease. *Arch. Neurol.* **59**(6), 955–962.

Chen, J. G., Charles, H. C., Barboriak, D. P., and Doraiswamy, P. M. (2000). Magnetic resonance spectroscopy in Alzheimer's disease: Focus on N-acetylaspartate. *Acta. Neurol. Scand. Suppl.* **176,** 20–26.

Cheng, L. L., Newell, K., Mallory, A. E., Hyman, B. T., and Gonzalez, R. G. (2002). Quantification of neurons in Alzheimer and control brains with *ex vivo* high resolution magic angle spinning proton magnetic resonance spectroscopy and stereology. *Magn. Reson. Imaging* **20**(7), 527–533.

Christiansen, P., Schlosser, A., and Henriksen, O. (1995). Reduced N-acetylaspartate content in the frontal part of the brain in patients with probable Alzheimer's disease. *Magn. Reson. Imaging* **13**(3), 457–462.

Cordato, N. J., Halliday, G. M., Harding, A. J., Hely, M. A., and Morris, J. G. (2000). Regional brain atrophy in progressive supranuclear palsy and Lewy body disease. *Ann. Neurol.* **47**(6), 718–728.

Coulthard, E., Firbank, M., English, P., Welch, J., Birchall, D., O'Brien, J., and Griffiths, T. D. (2006). Proton magnetic resonance spectroscopy in frontotemporal dementia. *J. Neurol.* **253**(7), 861–868.

Cummings, J. L., Vinters, H. V., Cole, G. M., and Khachaturian, Z. S. (1998). Alzheimer's disease: Etiologies, pathophysiology, cognitive reserve, and treatment opportunities. *Neurology* **51**(1 Suppl 1), S2–S17; discussion S65–17.

Dickson, D. (2000). Alzheimer-Parkinson disease overlap: Neuropathology. *In* "Neurodegenerative dementias" (C Clark and J Trojanowski, Eds.), pp. 247–259. McGraw-Hill, New York City.

Dickson, D. W. (1997). The pathogenesis of senile plaques. *J. Neuropathol. Exp. Neurol.* **56**(4), 321–339.

Dixon, R. M., Bradley, K. M., Budge, M. M., Styles, P., and Smith, A. D. (2002). Longitudinal quantitative proton magnetic resonance spectroscopy of the hippocampus in Alzheimer's disease. *Brain* **125**(Pt 10), 2332–2341.

Doraiswamy, P. M., Charles, H. C., and Krishnan, K. R. (1998). Prediction of cognitive decline in early Alzheimer's disease. *Lancet* **352**(9141), 1678.

Ernst, T., Chang, L., Melchor, R., and Mehringer, C. M. (1997). Frontotemporal dementia and early Alzheimer disease: Differentiation with frontal lobe H-1 MR spectroscopy. *Radiology* **203**(3), 829–836.

Fenili, D., Brown, M., Rappaport, R., and McLaurin, J. (2007). Properties of scyllo-inositol as a therapeutic treatment of AD-like pathology. *J. Mol. Med.* **85**(6), 603–611.

Firbank, M. J., Harrison, R. M., and O'Brien, J. T. (2002). A comprehensive review of proton magnetic resonance spectroscopy studies in dementia and Parkinson's disease. *Dement. Geriatr. Cogn. Disord.* **14**(2), 64–76.

Franczak, M., Prost, R. W., Antuono, P. G., Mark, L. P., Jones, J. L., and Ulmer, J. L. (2007). Proton magnetic resonance spectroscopy of the hippocampus in patients with mild cognitive impairment: A pilot study. *J. Comput. Assist. Tomogr.* **31**(5), 666–670.

Frederick, B., Satlin, A., Wald, L. L., Hennen, J., Bodick, N., and Renshaw, P. F. (2002). Brain proton magnetic resonance spectroscopy in Alzheimer disease: Changes after treatment with xanomeline. *Am. J. Geriatr. Psychiatry.* **10**(1), 81–88.

Frederick, B. B., Satlin, A., Yurgelun-Todd, D. A., and Renshaw, P. F. (1997). *In vivo* proton magnetic resonance spectroscopy of Alzheimer's disease in the parietal and temporal lobes. *Biol. Psychiatry.* **42**(2), 147–150.

Galvin, J. E., Pollack, J., and Morris, J. C. (2006). Clinical phenotype of Parkinson disease dementia. *Neurology* **67**(9), 1605–1611.

Ganguli, M., Dodge, H., Shen, C., and DeKosky, S. T. (2004). Mild cognitive impairment, amnestic type: An epidemiological study. *Neurology* **63,** 115–121.

Garcia Santos, J. M., Gavrila, D., Antunez, C., Tormo, M. J., Salmeron, D., Carles, R., Jimenez Veiga, J., Parrilla, G., Torres del Rio, S., Fortuna, L., and Navarro, C. (2008). Magnetic resonance spectroscopy performance for detection of dementia, Alzheimer's disease and mild cognitive impairment in a community-based survey. *Dement. Geriatr. Cogn. Disord.* **26**(1), 15–25.

Garrard, P., Schott, J. M., MacManus, D. G., Hodges, J. R., Fox, N. C., and Waldman, A. D. (2006). Posterior cingulate neurometabolite profiles and clinical phenotype in frontotemporal dementia. *Cogn. Behav. Neurol.* **19**(4), 185–189.

Godbolt, A. K., Waldman, A. D., MacManus, D. G., Schott, J. M., Frost, C., Cipolotti, L., Fox, N. C., and Rossor, M. N. (2006). MRS shows abnormalities before symptoms in familial Alzheimer disease. *Neurology* **66**(5), 718–722.

Gomez-Isla, T., Growdon, W. B., McNamara, M., Newell, K., Gomez-Tortosa, E., Hedley-Whyte, E. T., and Hyman, B. T. (1999). Clinicopathologic correlates in temporal cortex in dementia with Lewy bodies. *Neurology* **53**(9), 2003–2009.

Griffith, H. R., den Hollander, J. A., Okonkwo, O. C., O'Brien, T., Watts, R. L., and Marson, D. C. (2008a). Brain metabolism differs in Alzheimer's disease and Parkinson's disease dementia *Alzheimer's & Dementia,* In press.

Griffith, H. R., den Hollander, J. A., Okonkwo, O. C., O'Brien, T., Watts, R. L., and Marson, D. C. (2008b). Brain N-acetylaspartate is Reduced in Parkinson Disease With Dementia. *Alzheimer. Dis. Assoc. Disord.* **22**(1), 54–60.

Griffith, H. R., den Hollander, J. A., Stewart, C. C., Evanochko, W. T., Buchthal, S. D., Harrell, L. E., Zamrini, E. Y., Brockington, J. C., and Marson, D. C. (2007). Elevated brain scyllo-inositol concentrations in patients with Alzheimer's disease. *NMR Biomed* **20,** 709–716.

Griffith, H. R., Hollander, J. A., Okonkwo, O., Evanochko, W. T., Harrell, L. E., Zamrini, E. Y., Brockington, J. C., and Marson, D. C. (2007). Executive function is associated with brain proton magnetic resonance spectroscopy in amnestic mild cognitive impairment. *J. Clin. Exp. Neuropsychol.* **29**(6), 599–609.

Griffith, H. R., Netson, K. L., Harrell, L. E., Zamrini, E. Y., Brockington, J. C., and Marson, D. C. (2006). Amnestic mild cognitive impairment: Diagnostic outcomes and clinical prediction over a two-year time period. *J. Int. Neuropsychol. Soc.* **12**(2), 166–175.

Griffith, H. R., Okonkwo, O. C., den Hollander, J. A., Belue, K., Lanza, S., Harrell, L. E., Brockington, J. C., Clark, D. G., and Marson, D. C. (2007a). Brain proton MRS is correlated with financial abilities in patients with Alzheimer's disease. *Brain Imaging & Behavior* **1,** 23–29.

Griffith, H. R., Okonkwo, O. C., den Hollander, J. A., Belue, K., Lanza, S., Harrell, L. E., Brockington, J. C., Clark, D. G., and Marson, D. C. (2007b). Brain proton MRS is correlated with financial abilities in patients with Alzheimer's disease. *Brain Imaging & Behavior* **1,** 23–29.

Griffith, H. R., Okonkwo, O. C., O'Brien, T., and Hollander, J. A. (2008). Reduced brain glutamate in patients with Parkinson's disease. *NMR Biomed.* **21**(4), 381–387.

Griffith, H. R., Stewart, C. C., den Hollander, J. A., Evanochko, W. T., Buchthal, S. D., Harrell, L. E., Zamrini, E. Y., Brockington, J. C., and Marson, D. C. (2006). *In-vivo* 3T 1H Magnetic Resonance Spectroscopy of the Brain Reveals Elevated Scyllo-inositol in Patients with Mild Alzheimer's Disease. Paper presented at the Proceedings of the International Society of Magnetic Resonance in Medicine, Seattle, WA.

Hattori, N., Abe, K., Sakoda, S., and Sawada, T. (2002). Proton MR spectroscopic study at 3 Tesla on glutamate/glutamine in Alzheimer's disease. *Neuroreport* **13**(1), 183–186.

Herminghaus, S., Frolich, L., Gorriz, C., Pilatus, U., Dierks, T., Wittsack, H. J., Lanfermann, H., Maurer, K., and Zanella, F. E. (2003). Brain metabolism in Alzheimer disease and vascular dementia assessed by *in vivo* proton magnetic resonance spectroscopy. *Psychiatry. Res.* **123**(3), 183–190.

Heun, R., Schlegel, S., Graf-Morgenstern, M., Tintera, J., Gawehn, J., and Stoeter, P. (1997). Proton magnetic resonance spectroscopy in dementia of Alzheimer type. *Int. J. Geriatr. Psychiatry.* **12**(3), 349–358.

Holmes, C., Cairns, N., Lantos, P., and Mann, A. (1999). Validity of current clinical criteria for Alzheimer's disease, vascular dementia and dementia with Lewy bodies. *Br. J. Psychiatry* **174,** 45–50.

Jessen, F., Block, W., Traber, F., Keller, E., Flacke, S., Lamerichs, R., Schild, H. H., and Heun, R. (2001). Decrease of N-acetylaspartate in the MTL correlates with cognitive decline of AD patients. *Neurology* **57**(5), 930–932.

Jessen, F., Block, W., Traber, F., Keller, E., Flacke, S., Papassotiropoulos, A., Lamerichs, R., Heun, R., and Schild, H. H. (2000). Proton MR spectroscopy detects a relative decrease of N-acetylaspartate in the medial temporal lobe of patients with AD. *Neurology* **55**(5), 684–688.

Jessen, F., Traeber, F., Freymann, K., Maier, W., Schild, H. H., and Block, W. (2006). Treatment monitoring and response prediction with proton MR spectroscopy in AD. *Neurology* **67**(3), 528–530.

Jones, R. S., and Waldman, A. D. (2004). 1H-MRS evaluation of metabolism in Alzheimer's disease and vascular dementia. *Neurol. Res.* **26**(5), 488–495.

Kaiser, L. G., Schuff, N., Cashdollar, N., and Weiner, M. W. (2005). Scyllo-inositol in normal aging human brain: ^{1}H magnetic resonance spectroscopy study at 4 Tesla. *NMR. Biomed.* **18**(1), 51–55.

Kantarci, K., Jack, C. R., Jr., Xu, Y. C., Campeau, N. G., O'Brien, P. C., Smith, G. E., Ivnik, R. J., Boeve, B. F., Kokmen, E., Tangalos, E. G., and Petersen, R. C. (2000). Regional metabolic patterns in mild cognitive impairment and Alzheimer's disease: A ^{1}H MRS study. *Neurology* **55**(2), 210–217.

Kantarci, K., Knopman, D. S., Dickson, D. W., Parisi, J. E., Whitwell, J. L., Weigand, S. D., Josephs, K. A., Boeve, B. F., Petersen, R. C., and Jack, C. R., Jr. (2008). Alzheimer disease: Postmortem neuropathologic correlates of antemortem 1H MR spectroscopy metabolite measurements. *Radiology* **248**(1), 210–220.

Kantarci, K., Petersen, R. C., Boeve, B. F., Knopman, D. S., Tang-Wai, D. F., O'Brien, P. C., Weigand, S. D., Edland, S. D., Smith, G. E., Ivnik, R. J., Ferman, T. J., Tangalos, E. G., *et al.* (2004). 1H MR spectroscopy in common dementias. *Neurology* **63**(8), 1393–1398.

Kantarci, K., Reynolds, G., Petersen, R. C., Boeve, B. F., Knopman, D. S., Edland, S. D., Smith, G. E., Ivnik, R. J., Tangalos, E. G., and Jack, C. R., Jr. (2003). Proton MR spectroscopy in mild cognitive impairment and Alzheimer disease: Comparison of 1.5 and 3 T. *AJNR. Am. J. Neuroradiol.* **24**(5), 843–849.

Kantarci, K., Smith, G. E., Ivnik, R. J., Petersen, R. C., Boeve, B. F., Knopman, D. S., Tangalos, E. G., and Jack, C. R., Jr. (2002). ^{1}H magnetic resonance spectroscopy, cognitive function, and apolipoprotein E genotype in normal aging, mild cognitive impairment and Alzheimer's disease. *J. Int. Neuropsychol. Soc.* **8**(7), 934–942.

Kantarci, K., Weigand, S. D., Petersen, R. C., Boeve, B. F., Knopman, D. S., Gunter, J., Reyes, D., Shiung, M., O'Brien, P. C., Smith, G. E., Ivnik, R. J., Tangalos, E. G., *et al.* (2007). Longitudinal 1H MRS changes in mild cognitive impairment and Alzheimer's disease. *Neurobiol. Aging.* **28**(9), 1330–1339.

Kantarci, K., Xu, Y., Shiung, M. M., O'Brien, P. C., Cha, R. H., Smith, G. E., Ivnik, R. J., Boeve, B. F., Edland, S. D., Kokmen, E., Tangalos, E. G., Petersen, R. C., *et al.* (2002). Comparative diagnostic utility of different MR modalities in mild cognitive impairment and Alzheimer's disease. *Dement. Geriatr. Cogn. Disord.* **14**(4), 198–207.

Kario, K., Matsuo, T., Hoshide, S., Umeda, Y., and Shimada, K. (1999). Effect of thrombin inhibition in vascular dementia and silent cerebrovascular disease. An MR spectroscopy study. *Stroke* **30**(5), 1033–1037.

Kattapong, V. J., Brooks, W. M., Wesley, M. H., Kodituwakku, P. W., and Rosenberg, G. A. (1996). Proton magnetic resonance spectroscopy of vascular- and Alzheimer-type dementia. *Arch. Neurol.* **53**(7), 678–680.

Kickler, N., Krack, P., Fraix, V., Lebas, J. F., Lamalle, L., Durif, F., Krainik, A., Remy, C., Segebarth, C., and Pollak, P. (2007). Glutamate measurement in Parkinson's disease using MRS at 3 T field strength. *NMR. Biomed.*

Kizu, O., Yamada, K., Ito, H., and Nishimura, T. (2004). Posterior cingulate metabolic changes in frontotemporal lobar degeneration detected by magnetic resonance spectroscopy. *Neuroradiology* **46**(4), 277–281.

Klunk, W. E., Engler, H., Nordberg, A., Wang, Y., Blomqvist, G., Holt, D. P., Bergstrom, M., Savitcheva, I., Huang, G. F., Estrada, S., Ausen, B., Debnath, M. L., *et al.* (2004). Imaging brain amyloid in Alzheimer's disease with Pittsburgh Compound-B. *Ann. Neurol.* **55**(3), 306–319.

Klunk, W. E., Panchalingam, K., McClure, R. J., Stanley, J. A., and Pettegrew, J. W. (1998). Metabolic alterations in postmortem Alzheimer's disease brain are exaggerated by Apo-E4. *Neurobiol. Aging.* **19**(6), 511–515.

Kovari, E., Gold, G., Herrmann, F. R., Canuto, A., Hof, P. R., Bouras, C., and Giannakopoulos, P. (2003). Lewy body densities in the entorhinal and anterior cingulate cortex predict cognitive deficits in Parkinson's disease. *Acta. Neuropathol. (Berl)* **106**(1), 83–88.

Krishnan, K. R., Charles, H. C., Doraiswamy, P. M., Mintzer, J., Weisler, R., Yu, X., Perdomo, C., Ieni, J. R., and Rogers, S. (2003). Randomized, placebo-controlled trial of the effects of donepezil on neuronal markers and hippocampal volumes in Alzheimer's disease. *Am. J. Psychiatry.* **160**(11), 2003–2011.

Kwock, L. (1998). Localized MR spectroscopy: Basic principles. *Neuroimaging Clin. N. Am.* **8**(4), 713–731.

Lazeyras, F., Charles, H. C., Tupler, L. A., Erickson, R., Boyko, O. B., and Krishnan, K. R. (1998). Metabolic brain mapping in Alzheimer's disease using proton magnetic resonance spectroscopy. *Psychiatry. Res.* **82**(2), 95–106.

Litvan, I., Mohr, E., Williams, J., Gomez, C., and Chase, T. N. (1991). Differential memory and executive functions in demented patients with Parkinson's and Alzheimer's disease. *J. Neurol. Neurosurg. Psychiatry* **54**(1), 25–29.

MacKay, S., Ezekiel, F., Di Sclafani, V., Meyerhoff, D. J., Gerson, J., Norman, D., Fein, G., and Weiner, M. W. (1996). Alzheimer disease and subcortical ischemic vascular dementia: Evaluation by combining MR imaging segmentation and H-1 MR spectroscopic imaging. *Radiology* **198**(2), 537–545.

Magistretti, P. J., and Pellerin, L. (1999). Cellular mechanisms of brain energy metabolism and their relevance to functional brain imaging. *Philos. Trans. R. Soc. Lond. B. Biol. Sci.* **354**(1387), 1155–1163.

Markesbery, W. R., Schmitt, F. A., Kryscio, R. J., Davis, D. G., Smith, C. D., and Wekstein, D. R. (2006). Neuropathologic substrate of mild cognitive impairment. *Arch. Neurol.* **63**(1), 38–46.

Martinez-Bisbal, M. C., Arana, E., Marti-Bonmati, L., Molla, E., and Celda, B. (2004). Cognitive impairment: Classification by 1H magnetic resonance spectroscopy. *Eur J. Neurol.* **11**(3), 187–193.

McKeith, I. G., Dickson, D. W., Lowe, J., Emre, M., O'Brien, J. T., Feldman, H., Cummings, J., Duda, J. E., Lippa, C., Perry, E. K., Aarsland, D., Arai, H., *et al.* (2005). Diagnosis and management of dementia with Lewy bodies: Third report of the DLB Consortium. *Neurology* **65**(12), 1863–1872.

McLaurin, J., Golomb, R., Jurewicz, A., Antel, J. P., and Fraser, P. E. (2000). Inositol stereoisomers stabilize an oligomeric aggregate of Alzheimer amyloid beta peptide and inhibit abeta -induced toxicity. *J. Biol. Chem.* **275**(24), 18495–18502.

McMurtray, A., Clark, D. G., Christine, D., and Mendez, M. F. (2006). Early-onset dementia: frequency and causes compared to late-onset dementia. *Dement. Geriatr. Cogn. Disord.* **21**(2), 59–64.

Metastasio, A., Rinaldi, P., Tarducci, R., Mariani, E., Feliziani, F. T., Cherubini, A., Pelliccioli, G. P., Gobbi, G., Senin, U., and Mecocci, P. (2006). Conversion of MCI to dementia: Role of proton magnetic resonance spectroscopy. *Neurobiol. Aging.* **27**(7), 926–932.

Meyerhoff, D. J., MacKay, S., Constans, J. M., Norman, D., Van Dyke, C., Fein, G., and Weiner, M. W. (1994). Axonal injury and membrane alterations in Alzheimer's disease suggested by *in vivo* proton magnetic resonance spectroscopic imaging. *Ann. Neurol.* **36**(1), 40–47.

Mihara, M., Hattori, N., Abe, K., Sakoda, S., and Sawada, T. (2006). Magnetic resonance spectroscopic study of Alzheimer's disease and frontotemporal dementia/Pick complex. *Neuroreport* **17**(4), 413–416.

Moats, R. A., Ernst, T., Shonk, T. K., and Ross, B. D. (1994). Abnormal cerebral metabolite concentrations in patients with probable Alzheimer disease. *Magn. Reson. Med.* **32**(1), 110–115.

Modrego, P. J., Fayed, N., and Pina, M. A. (2005). Conversion from mild cognitive impairment to probable Alzheimer's disease predicted by brain magnetic resonance spectroscopy. *Am. J. Psychiatry.* **162**(4), 667–675.

Modrego, P. J., Pina, M. A., Fayed, N., and Diaz, M. (2006). Changes in metabolite ratios after treatment with rivastigmine in Alzheimer's disease: A nonrandomised controlled trial with magnetic resonance spectroscopy. *CNS. Drugs.* **20**(10), 867–877.

Molina, J. A., Garcia-Segura, J. M., Benito-Leon, J., Gomez-Escalonilla, C., del Ser, T., Martinez, V., and Viano, J. (2002). Proton magnetic resonance spectroscopy in dementia with Lewy bodies. *Eur. Neurol.* **48**(3), 158–163.

Mueller, S. G., Schuff, N., and Weiner, M. W. (2006). Evaluation of treatment effects in Alzheimer's and other neurodegenerative diseases by MRI and MRS. *NMR. Biomed.* **19**(6), 655–668.

Osaki, Y., Morita, Y., Fukumoto, M., Akagi, N., Yoshida, S., and Doi, Y. (2005). Three-dimensional stereotactic surface projection SPECT analysis in Parkinson's disease with and without dementia. *Mov. Disord.* **20**(8), 999–1005.

Parnetti, L., Lowenthal, D. T., Presciutti, O., Pelliccioli, G. P., Palumbo, R., Gobbi, G., Chiarini, P., Palumbo, B., Tarducci, R., and Senin, U. (1996). 1H-MRS, MRI-based hippocampal volumetry, and 99mTc-HMPAO-SPECT in normal aging, age-associated memory impairment, and probable Alzheimer's disease. *J. Am. Geriatr. Soc.* **44**(2), 133–138.

Parnetti, L., Tarducci, R., Presciutti, O., Lowenthal, D. T., Pippi, M., Palumbo, B., Gobbi, G., Pelliccioli, G. P., and Senin, U. (1997). Proton magnetic resonance spectroscopy can differentiate Alzheimer's disease from normal aging. *Mech. Ageing. Dev.* **97**(1), 9–14.

Petersen, R. C. (2000). Mild cognitive impairment: Transition between aging and Alzheimer's disease. *Neurologia* **15**(3), 93–101.

Petersen, R. C. (2004). Mild cognitive impairment as a diagnostic entity. *J. Intern. Med.* **256**(3), 183–194.

Petersen, R. C., Doody, R., Kurz, A., Mohs, R. C., Morris, J. C., Rabins, P. V., Ritchie, K., Rossor, M., Thal, L., and Winblad, B. (2001). Current concepts in mild cognitive impairment. *Arch. Neurol.* **58**(12), 1985–1992.

Petersen, R. C., Smith, G. E., Waring, S. C., Ivnik, R. J., Tangalos, E. G., and Kokmen, E. (1999). Mild cognitive impairment: Clinical characterization and outcome. *Arch. Neurol.* **56**(3), 303–308.

Pfefferbaum, A., Adalsteinsson, E., Spielman, D., Sullivan, E. V., and Lim, K. O. (1999). *In vivo* brain concentrations of N-acetyl compounds, creatine, and choline in Alzheimer disease. *Arch. Gen. Psychiatry.* **56**(2), 185–192.

Pletnikova, O., West, N., Lee, M. K., Rudow, G. L., Skolasky, R. L., Dawson, T. M., Marsh, L., and Troncoso, J. C. (2005). Abeta deposition is associated with enhanced cortical alpha-synuclein lesions in Lewy body diseases. *Neurobiol. Aging.* **26**(8), 1183–1192.

Rai, G. S., McConnell, J. R., Waldman, A., Grant, D., and Chaudry, M. (1999). Brain proton spectroscopy in dementia: An aid to clinical diagnosis. *Lancet.* **353**(9158), 1063–1064.

Rami, L., Caprile, C., Gómez-Ansón, B., Sánchez-Valle, R., Monte, G. C., Bosch, B., and Molinuevo, J. L. (2008). Naming is associated with left temporal pole metabolite levels in neurodegenerative diseases. *Dement. Geriatr. Cogn. Disord.* **25**(3), 212–217.

Rami, L., Gomez-Anson, B., Bosch, B., Sanchez-Valle, R., Monte, G. C., Villar, A., and Molinuevo, J. L. (2007). Cortical brain metabolism as measured by proton spectroscopy is related to memory performance in patients with amnestic mild cognitive impairment and Alzheimer's disease. *Dement. Geriatr. Cogn. Disord.* **24**(4), 274–279.

Rami, L., Gomez-Anson, B., Sanchez-Valle, R., Bosch, B., Monte, G. C., Llado, A., and Molinuevo, J. L. (2007). Longitudinal study of amnesic patients at high risk for Alzheimer's disease: Clinical, neuropsychological and magnetic resonance spectroscopy features. *Dement. Geriatr. Cogn. Disord.* **24**(5), 402–410.

Riederer, P., and Hoyer, S. (2006). From benefit to damage. Glutamate and advanced glycation end products in Alzheimer brain. *J. Neural. Transm.* **113**(11), 1671–1677.

Rose, S. E., de Zubicaray, G. I., Wang, D., Galloway, G. J., Chalk, J. B., Eagle, S. C., Semple, J., and Doddrell, D. M. (1999). A 1H MRS study of probable Alzheimer's disease and normal aging: Implications for longitudinal monitoring of dementia progression. *Magn. Reson. Imaging.* **17**(2), 291–299.

Rosen, Y., and Lenkinski, R. E. (2007). Recent advances in magnetic resonance neurospectroscopy. *Neurotherapeutics* **4**(3), 330–345.

Ross, B., and Bluml, S. (2001). Magnetic resonance spectroscopy of the human brain. *Anat. Rec.* **265**(2), 54–84.

Schubert, F., Gallinat, J., Seifert, F., and Rinneberg, H. (2004). Glutamate concentrations in human brain using single voxel proton magnetic resonance spectroscopy at 3 Tesla. *Neuroimage* **21**(4), 1762–1771.

Schuff, N., Amend, D., Ezekiel, F., Steinman, S. K., Tanabe, J., Norman, D., Jagust, W., Kramer, J. H., Mastrianni, J. A., Fein, G., and Weiner, M. W. (1997). Changes of hippocampal N-acetyl aspartate and volume in Alzheimer's disease. A proton MR spectroscopic imaging and MRI study. *Neurology* **49**(6), 1513–1521.

Shonk, T. K., Moats, R. A., Gifford, P., Michaelis, T., Mandigo, J. C., Izumi, J., and Ross, B. D. (1995). Probable Alzheimer disease: Diagnosis with proton MR spectroscopy. *Radiology* **195**(1), 65–72.

Storandt, M., Grant, E. A., Miller, J. P., and Morris, J. C. (2002). Rates of progression in mild cognitive impairment and early Alzheimer's disease. *Neurology* **59**(7), 1034–1041.

Strange, K., Emma, F., Paredes, A., and Morrison, R. (1994). Osmoregulatory changes in myo-inositol content and Na+/myo-inositol cotransport in rat cortical astrocytes. *Glia* **12**(1), 35–43.

Summerfield, C., Gomez-Anson, B., Tolosa, E., Mercader, J. M., Marti, M. J., Pastor, P., and Junque, C. (2002). Dementia in Parkinson disease: A proton magnetic resonance spectroscopy study. *Arch. Neurol.* **59**(9), 1415–1420.

Terry, R. D., Masliah, E., Salmon, D. P., Butters, N., DeTeresa, R., Hill, R., Hansen, L. A., and Katzman, R. (1991). Physical basis of cognitive alterations in Alzheimer's disease: Synapse loss is the major correlate of cognitive impairment. *Ann. Neurol.* **30**(4), 572–580.

Tierney, M. C., Szalai, J. P., Snow, W. G., Fisher, R. H., Nores, A., Nadon, G., Dunn, E., and St George-Hyslop, P. H. (1996). Prediction of probable Alzheimer's disease in memory-impaired patients: A prospective longitudinal study. *Neurology* **46**(3), 661–665.

Tiraboschi, P., Hansen, L. A., Alford, M., Sabbagh, M. N., Schoos, B., Masliah, E., Thal, L. J., and Corey-Bloom, J. (2000). Cholinergic dysfunction in diseases with Lewy bodies. *Neurology* **54**(2), 407–411.

Valenzuela, M. J., and Sachdev, P. (2001). Magnetic resonance spectroscopy in AD. *Neurology* **56**(5), 592–598.

Waldman, A. D., and Rai, G. S. (2003). The relationship between cognitive impairment and *in vivo* metabolite ratios in patients with clinical Alzheimer's disease and vascular dementia: A proton magnetic resonance spectroscopy study. *Neuroradiology* **45**(8), 507–512.

Waldman, A. D., Rai, G. S., McConnell, J. R., Chaudry, M., and Grant, D. (2002). Clinical brain proton magnetic resonance spectroscopy for management of Alzheimer's and sub-cortical ischemic vascular dementia in older people. *Arch. Gerontol. Geriatr.* **35**(2), 137–142.

Watanabe, T., Akiguchi, I., Yagi, H., Onishi, K., Kawasaki, T., Shiino, A., and Inubushi, T. (2002). Proton magnetic resonance spectroscopy and white matter hyperintensities on magnetic resonance imaging in patients with Alzheimer's disease. *Ann. N. Y. Acad. Sci.* **977,** 423–429.

Weiss, U., Bacher, R., Vonbank, H., Kemmler, G., Lingg, A., and Marksteiner, J. (2003). Cognitive impairment: Assessment with brain magnetic resonance imaging and proton magnetic resonance spectroscopy. *J. Clin. Psychiatry.* **64**(3), 235–242.

Wiederkehr, S., Simard, M., Fortin, C., and van Reekum, R. (2008). Comparability of the clinical diagnostic criteria for vascular dementia: A critical review. Part I. *J Neuropsychiatry. Clin. Neurosci.* **20**(2), 150–161.

Xuan, X., Ding, M., and Gong, X. (2008). Proton magnetic resonance spectroscopy detects a relative decrease of N-acetylaspartate in the hippocampus of patients with dementia with Lewy bodies. *J. Neuroimaging* **18**(2), 137–141.

APPLICATION OF PET IMAGING TO DIAGNOSIS OF ALZHEIMER'S DISEASE AND MILD COGNITIVE IMPAIRMENT[1]

James M. Noble*,†,‡ and Nikolaos Scarmeas†,‡

*Department of Neurology, Harlem Hospital Center, Columbia University College of Physicians and Surgeons, New York 10037, USA

†Department of Neurology, Columbia University Medical Center, New York 10032, USA

‡Gertrude H. Sergievsky Center, and the Taub Institute for Research on Alzheimer's Disease and the Aging Brain, Columbia University Medical Center, New York 10032, USA

Alzheimer disease (AD) is the most common type of dementia and will become increasingly prevalent with the growing elderly population. Despite established clinical diagnostic tools, the workup for dementia among primary caregivers can be complicated and specialist referral may not be readily available. A host of AD diagnostic tests has been proposed to aid in diagnosis, including functional neuroimaging such as positron emission tomography (PET). We review the basis for FDG-PET and PiB-PET, as well as available operating statistics. From this we advise scenarios for use of PET in primary settings and referral centers, approach to its interpretation, and outline a clinical prediction model based on findings.

[1]Disclosures: Neither Drs. Noble nor Scarmeas have any relevant conflicts of interest.

INTERNATIONAL REVIEW OF NEUROBIOLOGY, VOL. 84
DOI: 10.1016/S0074-7742(09)00407-3

I. Background

Dementia is a common disorder among the elderly, becomes more prevalent with advancing age, is typically medically refractory, reduces life expectancy, and diminishes quality of life for patients and their caregivers. Worldwide 600 million people are over the age of 60 and that number is expected to double by 2025 (UN, 2003). Of those over age 60, approximately 26 million people have dementia of any type (Wimo *et al.*, 2007), and as the population ages, this is projected increase to 80 million by the year 2040 (ADI, 2006). The costs of care are staggering, with a current annual worldwide health cost estimated at $300 billion (Wimo *et al.*, 2007).

Despite the high prevalence of dementia among the elderly, the diagnostic workup can be complicated, time consuming, and difficult to interpret, particularly in the primary care setting. In this manuscript, we will review the biologic basis of positron emission technology (PET) in the diagnostic workup of Alzheimer disease (AD) and its symptomatic prodrome, mild cognitive impairment (MCI), as well as outline current recommendations for use PET, and identify its main limitations to improve appropriate selection of patients.

A. Neurobiology of Alzheimer Disease

Abnormal accumulation of beta-amyloid (Aβ) peptide is widely believed to be the underlying mechanism of pathologic and clinical changes seen in AD (Cummings, 2004). Changes typically begin in the transentorhinal and entorhinal cortex as well as the hippocampus (Delacourte *et al.*, 1999), and thus earliest impairments occur with episodic memory deficits. Despite extensive neuropathologic changes, some patients may remain asymptomatic in a prodromal state of AD (Dubois *et al.*, 2007). Alternatively, some patients develop cognitive complaints and have objective cognitive impairment but no functional decline in level of activities of daily living, and these are classified as MCI (Petersen *et al.*, 2001). As disease progresses, other brain structures become involved, initially temporal cortex and later parietal and frontal association cortices, and finally primary motor and sensory cortices and the neocortex (Delacourte *et al.*, 1999). Symptomatically, patients progress from episodic memory loss to additional, progressive deficits in visuospatial and semantic abilities, mood and behavioral disorders, and an eventual vegetative state.

B. Epidemiology of Alzheimer Disease

Alzheimer disease (AD) is the most common type of dementia, representing 60–70% of all patients with dementia. The prevalence increases with age from approximately 3–5% for those under 75, 15–18% for those age 75–84 years, and

45% for those older than 85 years (Alzheimer's Association, 2008; Gurland *et al.*, 1999; Hebert *et al.*, 2003; Mayeux, 2003b). In the US alone, an estimated 5.2 million have AD in 2008 and this may increase to 13.2 million by 2050 (Alzheimer's Association, 2008). Sporadic AD represents 98% of all AD cases and is likely due to a complex interaction of environmental, vascular, and genetic risk factors (Mayeux, 2003a; Tang *et al.*, 1998).

The epidemiologic data for patients at risk for late onset AD are equally daunting. In the US alone, 25 million persons are over the age of 65 (18 million aged 65–74 years, 12 million aged 75–84 years, and 4 million aged 85 years and over) (He *et al.*, 2005). Before developing AD, many patients may transition through amnestic MCI (Petersen *et al.*, 2001), or memory complaints with commensurate objective neuropsychological test deficits, but without appreciable impact in general cognition or daily independent function.

MCI portends to a significantly increased annual risk of conversion to AD compared to others in their age group (Manly *et al.*, 2008; Petersen *et al.*, 2001). In a multiethnic urban American population, amnestic MCI (including single or multiple domain impairment) is prevalent in 3.8% of nondemented elderly aged 65–75 and in 6.3% of those over 75 years of age (Manly *et al.*, 2005). Overall, most studies suggest that MCI patients have an annual conversion rate (i.e., progression) to AD of approximately 10–15% per year (Morris *et al.*, 2001; Petersen *et al.*, 1999). Recently, age-stratified data from our center have suggested that of normal elderly individuals, the annual conversion rate from normal cognition to amnestic MCI is 1.1% of those aged 65–69 years, 2.1% aged 70–74, 2.2% aged 75–79, and 3.4% aged 80 years and above. Moreover, among those with amnestic MCI, the annual conversion rate to AD is 3.2% of those aged 65–69, 4.5% aged 70–74, 9.7% aged 75–79, and 11.1% aged 80 years and above (Manly *et al.*, 2008).

C. [^{18}F]2-Fluoro-2-Deoxy-d-Glucose

Positron emission tomography (PET) enables imaging of biological activity, and thus can identify abnormal biological activity in a host of diseases. Regional differences in brain glucose metabolism using PET have been explored in dementia, and [^{18}F]2-fluoro-2-deoxy-D-glucose (FDG-PET) has become the most widely investigated radioligand since the first descriptions of cerebral metabolism in dementia over two decades ago (Benson *et al.*, 1981; Farkas *et al.*, 1982; Friedland *et al.*, 1983; Phelps *et al.*, 1982). Glucose hypometabolism identified in FDG-PET is thought to represent two findings: (1) local decreases in synaptic activity or synaptic dysfunction in neurons affected by Alzheimer type pathological changes and (2) decreased synaptic activity in regions receiving projections from these primary, diseased neurons (Friedland *et al.*, 1985; Hoffman *et al.*, 2000; Matsuda, 2007; McGeer *et al.*, 1986a,b, 1990a,b; Mielke *et al.*, 1996).

Thus, abnormalities in regional cerebral metabolism are thought to reflect the pattern of neuropathologic development of AD with early prominent AD changes in the medial temporal cortex and its projections to cingulate and parieto-temporal association cortices (Brun and Englund, 1986); although, all of these structures may become pathologically involved with disease progression (Braak and Braak, 1991).

Similarly, patients with MCI, with many in the earliest stages of AD change, may only have glucose hypometabolism in the medial temporal cortex (de Leon *et al.*, 2001; Nestor *et al.*, 2003). MCI patients with additional glucose hypometabolism in the parietal association cortex may be at greater risk for subsequent conversion to AD (Chetelat *et al.*, 2003; Mosconi *et al.*, 2004).

Despite the relatively lengthy history of FDG-PET in dementia, its clinical utility was not fully appreciated until relatively recently. The last American Academy of Neurology Practice Parameter for the diagnosis of dementia was released in 2001, at a time prior to any reports of PET operating statistics from large studies. In that report, PET was thought to "have promise for use as an adjunct to clinical diagnosis," but further studies were required "to establish the value that it brings to diagnosis over and above a competent clinical diagnosis." (Knopman *et al.*, 2001). Since then, several studies have been reported including one later that year, which included 284 patients undergoing dementia workup, with 138 having neuropathological diagnosis (Silverman *et al.*, 2001). The authors used a typical AD pattern of parietal and temporal hypometabolism, with or without frontal involvement, with a sensitivity of 94% (91/97; 95% CI, 89–99%) and a specificity of 73% (30/41; 95% CI, 60–87%) relative to neuropathologic diagnosis (Silverman *et al.*, 2001); other studies have found similar values of specificity and sensitivity of PET for AD, depending on comparisons with clinical or pathologic definitions of disease (Hoffman *et al.*, 2000; Patwardhan *et al.*, 2004).

In an attempt to clarify the sensitivity and specificity of PET, one meta-analysis of articles published through 2003 found significant faults in the generalizability of these studies overall (Patwardhan *et al.*, 2004). Namely, the authors noted that control patients may have dissimilar medical comorbidities, and that most of the studies were performed in specialty or tertiary referral centers, and may operate quite differently when used in primary care settings. Despite these limitations their best estimate suggested the summary sensitivity of PET was 86% (95% CI: 76–93%) and summary specificity was 86% (95% CI: 72–93%) (Patwardhan *et al.*, 2004), and subsequent individual studies have found similar results (Jagust *et al.*, 2007). Furthermore, in each of these studies, we cannot know whether individuals with an AD PET pattern, but without either clinical or pathologic correlate, would have eventually developed clinical AD. Thus, these patients may underestimate the specificity of FDG-PET (fewer false positives).

D. Pittsburgh Compound B-PET

Recent approaches to PET imaging in dementia have attempted to obviate the examination of glucose metabolism and its inherent limitations and focus on radioligands which bind with amyloid in the brain. One of the most promising and well studied of these ligands used amyloid binding dye thioflavin-T derivative [*N*-methyl-^{11}C]-2-(4′-methylaminophenyl)-6-hydroxybenzothiazole, better known as Pittsburgh Compound B (PiB)-PET, which binds to amyloid plaques and amyloid fibrils, but not neurfibrillary tangles. The first report of PiB-PET in humans suggested PiB affinity is generally inversely correlated with glucose metabolism on FDG-PET (Klunk, 2004) (e.g., regions with greater amyloid burden may be identified in regions with relative preservation of glucose metabolism); the highest degree of amyloid binding was associated with prefrontal cortex, and to a lesser degree in parietal, occipital, and temporal cortex (Klunk *et al.*, 2004; Mintun *et al.*, 2006). Subsequent studies have also suggested that relatively normal prefrontal glucose metabolism on FDG-PET despite high prefrontal amyloid binding on PiB-PET has led some to hypothesize additional mechanisms than just amyloid underlying neuronal dysfunction in AD (Edison *et al.*, 2007).

PiB binding has been demonstrated to be highly prevalent among larger studies of AD patients, and to a lesser degree but with a similar pattern among MCI patients (Kemppainen *et al.*, 2007). PiB binding may be present in as high as 22% of elderly individuals with healthy aging, and cortical PiB is significantly associated with episodic memory test performance among MCI and normal aged patients (Pike *et al.*, 2007), and facial and word recognition in all subjects (Edison *et al.*, 2007). One study of cognitively normal elderly and MCI subjects found no within-group relation between learning and memory testing and PiB binding (Jack *et al.*, 2008). To date, the relatively limited number of studies preclude definitive prediction of progression from MCI or a normal state to clinical AD based on PiB-PET, but MCI patients with an AD PiB-PET pattern may be more likely to progress to AD based on one study with relatively brief follow-up (Forsberg *et al.*, 2008).

Data regarding the sensitivity and specificity of PiB-PET in the diagnosis of AD is somewhat limited. One study found increased PiB binding in 89% of patients with clinically probable AD (Edison *et al.*, 2007); another found 100% sensitivity, with specificity dependent upon the age of the patient, ranging from 73 to 96% (Ng *et al.*, 2007). As with FDG-PET, the cause for the relatively low specificity (high false positives for AD) in some subjects is unclear, but could be related to several reasons. First, presymptomatic individuals could lead to false positive findings; autopsied brains of cognitively normal elderly can have pathologic AD changes in 25–67% of subjects (Crystal *et al.*, 1988; Morris *et al.*, 1996; Mortimer *et al.*, 2003; MRC-CFAS, 2007). Second, PiB amyloid binding is found in other cerebral amyloidoses including cerebral amyloid angiopathy (CAA)

(Lockhart *et al.*, 2007). Although the data are limited, the CAA pattern of PiB-PET signal may be distinguishable from AD; CAA may have a greater degree of occipital amyloid identified on PiB-PET than AD patients (Johnson *et al.*, 2007).

II. Application of PET to Primary and Specialty Care Settings

Given the data from these studies outlined above, it is worth considering the application and utility of PET in clinical practice. Prior to outlining the approach to its use, the current typical sequence leading to the workup and diagnosis of dementia bears review, to contextualize the potential role of PET within the diagnostic workup.

A. Diagnostic Workup of Dementia in Primary Care

For the workup of dementia patients, current recommendations for primary care physicians are to provide cognitive screens to only those with reported or suspected cognitive decline, or in the oldest old (Brodaty *et al.*, 1998; US Preventive Services Task Force, 2003). Without disease modifying treatment, the current ratio of benefit to harm of screening for dementia is unclear, and thus not recommended unless dementia is suspected (US Preventive Services Task Force, 2003). These practices may delay recognition relative to community screening programs (Barker *et al.*, 2005), but expanding current screening practices may be impractical for a busy primary care physicians, already conscious of cost concerns (Geldmacher, 2002). When concern for progressive cognitive impairment is raised in the primary care setting, one of several brief cognitive screening tests for dementia is often used. The best-studied cognitive screening tool is the Folstein Mini-Mental Status Examination (MMSE) (Folstein *et al.*, 1975; US Preventive Services Task Force, 2003), but scoring and interpretation may be somewhat dependent on education and age. If cognitive impairment is identified, other non-cognitive screening tests are recommend for the purposes of excluding uncommon causes non-degenerative cognitive decline (Knopman *et al.*, 2001). These tests include MRI (to rule out space occupying lesions, subclinical cerebral infarcts, and other conditions) and laboratory evaluations of thyroid function, vitamin B12 level, and syphilis serology. These tests have few limitations, are reproducible, and are relatively simple to perform and interpret by non-specialists.

Depending on individual practice, some primary care physicians may arrest the workup at this point and begin treatment. Others may refer to a specialist after screening, or even sooner—at the time of suspected cognitive impairment but prior to screening laboratory evaluation. The current structure of diagnostic workup is likely impractical for every patient with probable AD to be seen by a specialist. Although data are limited, US community primary care physicians refer 44% of their patients to specialists for diagnosis or confirmation (Fortinsky *et al.*, 1995). In comparison, more than 80% of community primary care physicians in Germany and academic primary care physicians in Canada make specialist referrals, most often neurologists or physicians specializing in both psychiatry and neurology (Pimlott *et al.*, 2006; Riedel-Heller *et al.*, 1999). Potential dementia specialists include neurologists (on average, 4.5 neurologists serve 100,000 US population) (World Health Organization, 2005), gerontologists (2.4 per 100,000), and geriatric psychiatrists (0.5 per 100,000) (Wimo *et al.*, 2007); behavioral neurologists and neuropsychiatrists comprise a small minority of general psychiatrists and neurologists. Thus, any increase in referrals to specialists would likely overwhelm the current system.

The statistical rigor and cost-effectiveness of a non-selective community-based screen using most of these diagnostic tools has not been widely reported. It has been argued that FDG-PET can be a cost saving measure for the diagnosis of AD in geriatric populations, in comparison to a conventional workup of clinical evaluation and exclusion of potential mimicking diseases (Silverman *et al.*, 2002). However, this study considered the population of having 51% prevalent disease (Silverman *et al.*, 2002), which would only be applicable to a highly selected, screened population within a primary caregiver's office, or would require an unselected population with all patients over 90 years of age, based on current epidemiologic estimates.

B. Diagnostic Workup of Dementia in Specialty Care Settings

In contrast to screening tests used largely for exclusionary purposes, confirmatory tests used in the diagnostic workup of patients with dementia are typically used as sensitive measures to rule in a diagnosis in a patient with moderate to highly probable disease. These tests include initial clinical evaluation, often by a specialist, with supportive testing, decided on a case by case basis, including neuropsychological testing, cerebrospinal fluid markers, and functional imaging such as FDG-PET. Each of these is costly, time consuming and requires a large amount of manpower to complete (McMahon *et al.*, 2003). As noted elsewhere, the additional diagnostic benefit of these diagnostic tests when done in expert settings is unclear, and these tests may have greater potential diagnostic usefulness when the diagnosis is less certain, as may be the case in primary care settings (Knopman, 2001).

III. Appropriate Use of PET

Before delving into the statistics of available studies regarding PET, the use of sensitivity, specificity, and related statistical tools also bears brief review. From the Table I, *Sensitivity* = A/(A + C) is the probability of having a positive test result among all those with disease. The complement of sensitivity is the *false negative proportion* = C/(A + C). *Specificity* = D/(B + D) is the probability of having a negative test result among those without disease. Similarly, the complement of specificity is the *false positive proportion* = B/(B + D). Drawn from Table I are terms that are often cited to guide clinicians' use for practical matters, and perhaps incorrectly so: the predictive value terms. The *positive predictive value* = A/(A + B) is the probability of having the disease given a positive test result. Similarly, the *negative predictive value* = D/(C + D) is the probability of not having the disease given a negative test result. These predictive values are often applied incorrectly; their usefulness is dependent upon the prevalence of disease within a given population, which may be unknown to the clinician. Furthermore, if data are derived from case-control studies, the predictive value tools cannot be used unless an equal sampling fraction of controls is established, and this is often difficult.

In place of these predictive value tools, some have suggested using likelihood ratio statistics (Grimes and Schulz, 2005), but these have not become as widely reported, despite their relative ease of using an appropriate table or computer program. The benefit of the likelihood ratios is that they incorporate statistics regarding all disease states and test results, without depending on prevalence. Instead, using Bayesian statistics these ratios allow the user to apply a likelihood ratio to a pretest probability (either according to known disease prevalence or clinical probability based on additional data) to derive a posttest probability. Thus, again from Table I, the *Likelihood ratio for disease if test positive* = sensitivity/(1-specificity) and the *Likelihood ratio for disease if test negative* = (1-sensitivity)/specificity. One can also consider the likelihood ratio positive to be the ratio of true positives to false positives, and the likelihood ratio negative as the ratio of false negatives to true negatives. The likelihood ratios in and of themselves have little inherent meaning but become relevant when applying them to individual patients.

TABLE I
THE 2 × 2 TABLE

	Disease	No disease
Positive test result	A	B
Negative test result	C	D

Although the likelihood ratio calculations are relatively straightforward, their application can be exacted by using Bayesian statistics, but these can be perceived as cumbersome. An alternative strategy is the use of the likelihood ratio table (Fig. 1), or perhaps more simply by remembering a few important numbers (Grimes and Schulz, 2005). Here we present all three methods of practical application of likelihood ratios.

A. Using Likelihood Ratio Tables

When considering the FDG-PET meta-analysis outlined above (Patwardhan *et al.*, 2004), which identified the composite sensitivity = 86% and specificity = 86% for AD, then the Likelihood ratio positive = 0.86/(1−0.86) = 6.1 and the Likelihood ratio negative = (1−0.86)/0.86 = 0.16. Once these have been calculated, we then look to Fig. 1A, a typical likelihood ratio table. To use the likelihood ratio table, begin with the pretest probability at the left. Using a straightedge, tracing from the pretest probability through the likelihood ratio will derive the posttest probability. Presumably, a posttest probability will need to be sufficiently high or low to make a clinical decision. Arguably, given that tertiary referral centers may be correct in clinical AD diagnosis approximately 90%, then any posttest probability at or above this would be satisfactory; a suitable threshold value for a low likelihood posttest probability can be determined on a case by case basis but presumably should be less than 5%.

Consider an example of a 50% pretest probability, which could be prevalence of AD in a nonagenarian population (or alternatively a primary physician's clinical suspicion based on interview alone) and thus the pretest probability without further information or diagnostic tests. In this scenario, a positive FDG-PET with likelihood ratio positive = 6.1 would increase the posttest probability to about 90%, while a negative test result with likelihood ratio negative = 0.16 would decrease the posttest probability to 10%. Alternatively, we could consider a single, larger study of PET in AD as outlined above (Silverman *et al.*, 2001) which identified sensitivity = 94% and specificity = 73%, and thus a likelihood ratio positive = 3.5 and likelihood ratio negative = 0.08. Beginning with a 50% pretest probability, our posttest probability with a positive test increases to only about 70% while a negative test decreases the probability to about 5%.

Exploring the likelihood ratio table, one can clearly see that an uncertain pretest probability is the best scenario for use of a diagnostic tool with sufficiently high likelihood ratio positive (ideally over 10) or low likelihood ratio negative (ideally 0.1 or less). Very high or low pretest probabilities will not be appreciably affected, regardless of the likelihood ratio test characteristics (Grimes and Schulz, 2005).

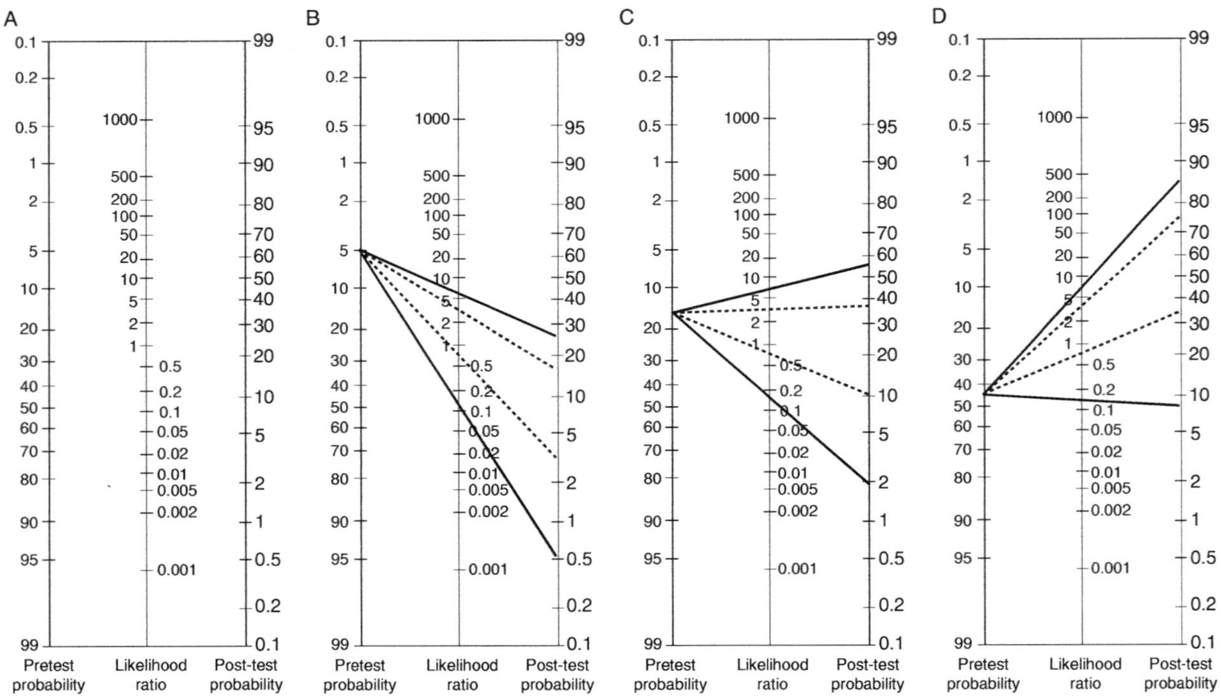

FIG. 1. Likelihood ratio tables of FDG-PET in the diagnosis of Alzheimer's disease in several age-stratified scenarios. (A) The likelihood ratio table. (B-D) Likelihood ratio tables for age stratified elderly populations, with pretest probability using only age-based prevalence statistics (Alzheimer's Association, 2008; Gurland *et al.*, 1999; Hebert *et al.*, 2003; Mayeux, 2003b). (B) Age less than 75 years (3–5% AD prevalence). (C) Age 75–84 years (15–18% AD prevalence). (D) Age 85 years and above (45% AD prevalence). Solid line: FDG-PET with sensitivity = 86% and specificity = 86% (Patwardhan *et al.*, 2004); (LR + = 6.1 LR− = 0.16). Dotted line: FDG-PET with sensitivity = 94% and specificity = 73% (Silverman *et al.*, 2001); LR + = 3.5 and LR− = 0.08. Figure (A) reprinted with permission from Elsevier (Grimes and Schulz, 2005).

Finally, in the Fig. 1B-D, we have applied the likelihood ratios derived from two studies of FDG-PET statistics (Patwardhan *et al.*, 2004; Silverman *et al.*, 2001) to visually demonstrate the utility of applying likelihood in several different scenarios. In this figure, we derived the pretest probability in each Fig. 1B-D from age-stratified prevalence statistics of AD (Alzheimer's Association, 2008; Gurland *et al.*, 1999; Hebert *et al.*, 2003; Mayeux *et al.*, 2003b) in an effort to demonstrate the utility of FDG-PET without additional screening or further workup. Clearly, the degree of certainty in diagnosis does not appreciably improve from this single test until applied to those older than 85 years of age.

B. Using a Likelihood Ratio "Mnemonic"

Should a likelihood ratio table not be immediately available, and if one chooses not to pursue the potentially difficult calculation steps outlined below, a simpler way has been suggested, requiring the user to remember a few numbers (Grimes and Schulz, 2005). Positive likelihood ratios 2, 5, and 10, approximately increase the posttest probability by 15% increments individually. That is, a likelihood ratio positive of 2 increases the posttest probability by an absolute difference of 15%, 5 by 30%, and 10 by 45%. Conversely, a likelihood ratio negative of 1/2 (0.5) decreases the posttest probability by an absolute difference of 15%, 1/5 (0.2) by 30%, and 1/10 (0.1) by 45%. Using this method considering one of the examples above, a likelihood ratio positive of 6.1 is similar to 5, which would increase a pretest probability of 50% absolutely by 30% to 80%. Similarly, a likelihood ratio negative of 0.08 is similar to 1/10 which would decrease a pretest probability of 50% absolutely by 45%, yielding a posttest probability of 5%, as estimated by the likelihood ratio tables. Overall, comparing this simplified methodology to the results above using the likelihood ratio tables, yields similar but less precise results.

C. Using Bayesian Calculations

Aside from using likelihood ratio tables or the mnemonic as suggested above, one can determine the exact posttest probabilities through a series of calculations, bearing in mind that pretest probabilities may be imprecise and based on clinical judgment. These calculations are

1. Calculate the pretest odds = pretest probability /(1-pretest probability)
2. Posttest odds = Pretest odds × likelihood ratio
3. Posttest probability = Posttest odds/(posttest odds +1)

Returning to the 50% pretest example using sensitivity and specificity for PET derived from the meta-analysis (Patwardhan *et al.*, 2004), for a positive test

Pretest odds = 0.50/(1−0.50) = 1 ("1:1 odds")
Posttest odds = 1 × 6.1 = 6.1
Posttest probability = 6.1/(6.1 + 1) = 86%

For a negative test

Posttest odds = 1 × 0.16 = 0.16
Posttest probability = 0.16/(1 + 0.16) = 14%

From this example we can see that the likelihood ratio table, menomnic, and calculations yield similar results for a test with 86% sensitivity and specificity, regarding a positive test result (posttest probability using the likelihood ratio table, 90%; mnemonic, 80%; calculation, 86%) and a negative test result (posttest probability using the likelihood ratio table, 10%; mnemonic, 5%; calculation, 14%).

D. Conclusions from Using the Likelihood Ratios for PET and AD

Based on these analyses, FDG-PET should be used only in a highly selected group of patients. Presumably, one would pursue a diagnostic workup until such a time that there is a high degree of certainty of either dementia or normal cognition. At this point FDG-PET should be reserved for patients having either (a) undergone an otherwise exhaustive diagnostic workup with a specialist in cognitive disorders or (b) partial workup with a primary care physician after having identified a cognitive disorder, but both yielding a still-uncertain diagnosis. This approach has been justified in expert centers attempting to distinguish ambiguous clinical cases of AD and frontotemporal dementia; FDG-PET changed the diagnosis in 26% of cases in which the examiner was uncertain or not completely confident in the diagnosis, versus 5% change of clinically certain cases after additional information provided by FDG-PET (Foster *et al.*, 2007). Current reimbursement practices in the United States reflect just such application of FDG-PET, and its use is restricted to consideration of a differential diagnosis of AD versus other neurodegenerative illnesses.

At this point, available data for either FDG-PET or PiB-PET in MCI are limited and preclude accurate derivation of a likelihood ratio test statistics. Neither study is currently indicated for suspected MCI patients, either as a diagnostic tool or as a prediction tool for progression. Should future studies reveal additional data regarding specificity and sensitivity of these imaging modalities, or any other MCI/AD diagnostic tool for that matter, one can apply such data to this formula and recapitulate this argument.

IV. Summary

Despite a relatively lengthy history of FDG-PET in the evaluation of dementia, it should still be used judiciously, and only in patients with a relatively uncertain cognitive diagnosis despite initial workup and evaluation, or if the etiology of their dementia is unclear. The use of FDG-PET as a blind screening tool is not recommended; although, screening in elderly at greatest risk for dementia (age > 90) is worth considering. Newer PET radioligands such as amyloid binding tools (PiB among them) show promise as a future diagnostic tool, but are not commercially available at this time and cannot be recommended for routine use based on available data. FDG-PET, PiB-PET, and other supportive diagnostic tests may have greater roles in diagnosis and monitoring disease should be an era of disease-modifying therapies and prevention strategies for AD develop.

References

ADI (2006). Common questions. In http://www.alz.co.uk/alzheimers/faq.html, Accessed December 11, 2006.

Alzheimer's Association (2008). *Alzheimer's Disease Facts and Figures.*

Barker, W. W., Luis, C., Harwood, D., Loewenstein, D., Bravo, M., Ownby, R., and Duara, R. (2005). The effect of a memory screening program on the early diagnosis of Alzheimer disease. *Alzheimers Dis. Assoc. Disord.* **19**(1), 1–7.

Benson, D. F., Kuhl, D. E., Phelps, M. E., Cummings, J. L., and Tsai, S. Y. (1981). Positron emission computed tomography in the diagnosis of dementia. *Trans. Am. Neurol. Assoc.* **106,** 68–71.

Braak, H., and Braak, E. (1991). Neuropathological stageing of Alzheimer-related changes. *Acta Neuropathol.* **82**(4), 239–259.

Brodaty, H., Clarke, J., Ganguli, M., Grek, A., Jorm, A. F., Khachaturian, Z., and Scherr, P. (1998). Screening for cognitive impairment in general practice: Toward a consensus. *Alzheimers Dis. Assoc. Disord.* **12**(1), 1–13.

Brun, A., and Englund, E. (1986). Brain changes in dementia of Alzheimer's type relevant to new imaging diagnostic methods. *Prog. Neuropsychopharmacol. Biol. Psychiatry* **10**(3–5), 297–308.

Chetelat, G., Desgranges, B., de la Sayette, V., Viader, F., Eustache, F., and Baron, J. C. (2003). Mild cognitive impairment: Can FDG-PET predict who is to rapidly convert to Alzheimer's disease? *Neurology* **60**(8), 1374–1377.

Crystal, H., Dickson, D., Fuld, P., Masur, D., Scott, R., Mehler, M., Masdeu, J., Kawas, C., Aronson, M., and Wolfson, L. (1988). Clinico-pathologic studies in dementia: Nondemented subjects with pathologically confirmed Alzheimer's disease. *Neurology* **38**(11), 1682–1687.

Cummings, J. L. (2004). Alzheimer's disease. *N. Engl. J. Med.* **351**(1), 56–67.

Delacourte, A., David, J. P., Sergeant, N., Buee, L., Wattez, A., Vermersch, P., Ghozali, F., Fallet-Bianco, C., Pasquier, F., Lebert, F., Petit, H., and Di Menza, C. (1999). The biochemical pathway of neurofibrillary degeneration in aging and Alzheimer's disease. *Neurology* **52**(6), 1158–1165.

de Leon, M. J., Convit, A., Wolf, O. T., Tarshish, C. Y., DeSanti, S., Rusinek, H., Tsui, W., Kandil, E., Scherer, A. J., Roche, A., Imossi, A., Thorn, E., *et al.* (2001). Prediction of cognitive decline in normal elderly subjects with 2-[(18)F]fluoro-2-deoxy-D-glucose/poitron-emission tomography (FDG/PET). *Proc. Natl. Acad. Sci. USA* **98**(19), 10966–10971.

Dubois, B., Feldman, H. H., Jacova, C., Dekosky, S. T., Barberger-Gateau, P., Cummings, J., Delacourte, A., Galasko, D., Gauthier, S., Jicha, G., Meguro, K., O'Brien, J., *et al.* (2007). Research criteria for the diagnosis of Alzheimer's disease: Revising the NINCDS-ADRDA criteria. *Lancet Neurol.* **6**(8), 734–746.

Edison, P., Archer, H. A., Hinz, R., Hammers, A., Pavese, N., Tai, Y. F., Hotton, G., Cutler, D., Fox, N., Kennedy, A., Rossor, M., and Brooks, D. J. (2007). Amyloid, hypometabolism, and cognition in Alzheimer disease: An [11C]PIB and [18F]FDG PET study. *Neurology* **68**(7), 501–508.

Farkas, T., Ferris, S. H., Wolf, A. P., De Leon, M. J., Christman, D. R., Reisberg, B., Alavi, A., Fowler, J. S., George, A. E., and Reivich, M. (1982). 18F-2-deoxy-2-fluoro-D-glucose as a tracer in the positron emission tomographic study of senile dementia. *Am. J. Psychiatry* **139**(3), 352–353.

Folstein, M. F., Folstein, S. E., and McHugh, P. R. (1975). "Mini-mental state". A practical method for grading the cognitive state of patients for the clinician. *J. Psychiatr. Res.* **12**(3), 189–198.

Forsberg, A., Engler, H., Almkvist, O., Blomquist, G., Hagman, G., Wall, A., Ringheim, A., Langstrom, B., and Nordberg, A. (2008). PET imaging of amyloid deposition in patients with mild cognitive impairment. *Neurobiol. Aging* **29,** 1456–1465.

Fortinsky, R. H., Leighton, A., and Wasson, J. H. (1995). Primary care physicians' diagnostic, management, and referral practices for older persons and families affected by dementia. *Res. Aging* **17,** 124–148.

Foster, N. L., Heidebrink, J. L., Clark, C. M., Jagust, W. J., Arnold, S. E., Barbas, N. R., DeCarli, C. S., Turner, R. S., Koeppe, R. A., Higdon, R., and Minoshima, S. (2007). FDG-PET improves accuracy in distinguishing frontotemporal dementia and Alzheimer's disease. *Brain* **130**(Pt 10), 2616–2635.

Friedland, R. P., Brun, A., and Budinger, T. F. (1985). Pathological and positron emission tomographic correlations in Alzheimer's disease. *Lancet* **1**(8422), 228.

Friedland, R. P., Budinger, T. F., Ganz, E., Yano, Y., Mathis, C. A., Koss, B., Ober, B. A., Huesman, R. H., and Derenzo, S. E. (1983). Regional cerebral metabolic alterations in dementia of the Alzheimer type: Positron emission tomography with [18F]fluorodeoxyglucose. *J. Comput. Assist. Tomogr.* **7**(4), 590–598.

Geldmacher, D. S. (2002). Cost-effective recognition and diagnosis of dementia. *Semin. Neurol.* **22**(1), 63–70.

Grimes, D. A., and Schulz, K. F. (2005). Refining clinical diagnosis with likelihood ratios. *Lancet* **365**(9469), 1500–1505.

Gurland, B. J., Wilder, D. E., Lantigua, R., Stern, Y., Chen, J., Killeffer, E. H., and Mayeux, R. (1999). Rates of dementia in three ethnoracial groups. *Int. J. Geriatr. Psychiatry* **14**(6), 481–493.

He, W., Sengupta, M., Velkoff, V. A., and DeBarros, K. (2005). 65+ in the United States: 2005 U.S. Department of Health and Human Services, National Institutes of Health, National Institute on Aging and the U.S. Department of Commerce, Economics and Statistics Administration, U.S. Census Bureau.

Hebert, L. E., Scherr, P. A., Bienias, J. L., Bennett, D. A., and Evans, D. A. (2003). Alzheimer disease in the US population: Prevalence estimates using the 2000 census. *Arch. Neurol.* **60**(8), 1119–1122.

Hoffman, J. M., Welsh-Bohmer, K. A., Hanson, M., Crain, B., Hulette, C., Earl, N., and Coleman, R. E. (2000). FDG PET imaging in patients with pathologically verified dementia. *J. Nucl. Med.* **41**(11), 1920–1928.

Jack, C. R., Jr., Lowe, V. J., Senjem, M. L., Weigand, S. D., Kemp, B. J., Shiung, M. M., Knopman, D. S., Boeve, B. F., Klunk, W. E., Mathis, C. A., and Petersen, R. C. (2008).

11C PiB and structural MRI provide complementary information in imaging of Alzheimer's disease and amnestic mild cognitive impairment. *Brain* **131**(Pt 3), 665–680.

Jagust, W., Reed, B., Mungas, D., Ellis, W., and Decarli, C. (2007). What does fluorodeoxyglucose PET imaging add to a clinical diagnosis of dementia? *Neurology* **69**(9), 871–877.

Johnson, K. A., Gregas, M., Becker, J. A., Kinnecom, C., Salat, D. H., Moran, E. K., Smith, E. E., Rosand, J., Rentz, D. M., Klunk, W. E., Mathis, C. A., Price, J. C., *et al.* (2007). Imaging of amyloid burden and distribution in cerebral amyloid angiopathy. *Ann. Neurol.* **62**(3), 229–234.

Kemppainen, N. M., Aalto, S., Wilson, I. A., Nagren, K., Helin, S., Bruck, A., Oikonen, V., Kailajarvi, M., Scheinin, M., Viitanen, M., Parkkola, R., and Rinne, J. O. (2007). PET amyloid ligand [11C]PIB uptake is increased in mild cognitive impairment. *Neurology* **68**(19), 1603–1606.

Klunk, W. E., Engler, H., Nordberg, A., Wang, Y., Blomqvist, G., Holt, D. P., Bergstrom, M., Savitcheva, I., Huang, G. F., Estrada, S., Ausen, B., Debnath, M. L., *et al.* (2004). Imaging brain amyloid in Alzheimer's disease with Pittsburgh Compound-B. *Ann. Neurol.* **55**(3), 306–319.

Knopman, D. (2001). Cerebrospinal fluid beta-amyloid and tau proteins for the diagnosis of Alzheimer disease. *Arch. Neurol.* **58**(3), 349–350.

Knopman, D. S., DeKosky, S. T., Cummings, J. L., Chui, H., Corey-Bloom, J., Relkin, N., Small, G. W., Miller, B., and Stevens, J. C. (2001). Practice parameter: Diagnosis of dementia (an evidence-based review). Report of the Quality Standards Subcommittee of the American Academy of Neurology. *Neurology* **56**(9), 1143–1153.

Lockhart, A., Lamb, J. R., Osredkar, T., Sue, L. I., Joyce, J. N., Ye, L., Libri, V., Leppert, D., and Beach, T. G. (2007). PIB is a non-specific imaging marker of amyloid-beta (Abeta) peptide-related cerebral amyloidosis. *Brain* **130**(Pt 10), 2607–2615.

Manly, J. J., Bell-McGinty, S., Tang, M. X., Schupf, N., Stern, Y., and Mayeux, R. (2005). Implementing diagnostic criteria and estimating frequency of mild cognitive impairment in an urban community. *Arch. Neurol.* **62**(11), 1739–1746.

Manly, J. J., Tang, M. X., Schupf, N., Stern, Y., Vonsattel, J. P., and Mayeux, R. (2008). Frequency and course of mild cognitive impairment in a multiethnic community. *Ann. Neurol.* **63,** 494–506.

Matsuda, H. (2007). The role of neuroimaging in mild cognitive impairment. *Neuropathology* **27**(6), 570–577.

Mayeux, R. (2003a). Apolipoprotein E, Alzheimer disease, and African Americans. *Arch. Neurol.* **60**(2), 161–163.

Mayeux, R. (2003b). Epidemiology of neurodegeneration. *Annu. Rev. Neurosci.* **26,** 81–104.

McGeer, P. L., Kamo, H., Harrop, R., McGeer, E. G., Martin, W. R., Pate, B. D., and Li, D. K. (1986a). Comparison of PET, MRI, and CT with pathology in a proven case of Alzheimer's disease. *Neurology* **36**(12), 1569–1574.

McGeer, P. L., Kamo, H., Harrop, R., Li, D. K., Tuokko, H., McGeer, E. G., Adam, M. J., Ammann, W., Beattie, B. L., Calne, D. B., *et al.* (1986b). Positron emission tomography in patients with clinically diagnosed Alzheimer's disease. *CMAJ* **134**(6), 597–607.

McGeer, E. G., McGeer, P. L., Harrop, R., Akiyama, H., and Kamo, H. (1990a). Correlations of regional postmortem enzyme activities with premortem local glucose metabolic rates in Alzheimer's disease. *J. Neurosci. Res.* **27**(4), 612–619.

McGeer, E. G., Peppard, R. P., McGeer, P. L., Tuokko, H., Crockett, D., Parks, R., Akiyama, H., Calne, D. B., Beattie, B. L., and Harrop, R. (1990b). 18Fluorodeoxyglucose positron emission tomography studies in presumed Alzheimer cases, including 13 serial scans. *Can. J. Neurol. Sci.* **17**(1), 1–11.

McMahon, P. M., Araki, S. S., Sandberg, E. A., Neumann, P. J., and Gazelle, G. S. (2003). Cost-effectiveness of PET in the diagnosis of Alzheimer disease. *Radiology* **228**(2), 515–522.

Mielke, R., Schroder, R., Fink, G. R., Kessler, J., Herholz, K., and Heiss, W. D. (1996). Regional cerebral glucose metabolism and postmortem pathology in Alzheimer's disease. *Acta Neuropathol.* **91**(2), 174–179.

Mintun, M. A., Larossa, G. N., Sheline, Y. I., Dence, C. S., Lee, S. Y., Mach, R. H., Klunk, W. E., Mathis, C. A., DeKosky, S. T., and Morris, J. C. (2006). [11C]PIB in a nondemented population: Potential antecedent marker of Alzheimer disease. *Neurology* **67**(3), 446–452.

Morris, J. C., Storandt, M., McKeel, D. W., Jr., Rubin, E. H., Price, J. L., Grant, E. A., and Berg, L. (1996). Cerebral amyloid deposition and diffuse plaques in "normal" aging: Evidence for presymptomatic and very mild Alzheimer's disease. *Neurology* **46**(3), 707–719.

Morris, J. C., Storandt, M., Miller, J. P., McKeel, D. W., Price, J. L., Rubin, E. H., and Berg, L. (2001). Mild cognitive impairment represents early-stage Alzheimer disease. *Arch. Neurol.* **58**(3), 397–405.

Mortimer, J. A., Snowdon, D. A., and Markesbery, W. R. (2003). Head circumference, education and risk of dementia: Findings from the Nun Study. *J. Clin. Exp. Neuropsychol.* **25**(5), 671–679.

Mosconi, L., Perani, D., Sorbi, S., Herholz, K., Nacmias, B., Holthoff, V., Salmon, E., Baron, J. C., De Cristofaro, M. T., Padovani, A., Borroni, B., Franceschi, M., *et al.* (2004). MCI conversion to dementia and the APOE genotype: A prediction study with FDG-PET. *Neurology* **63**(12), 2332–2340.

Nestor, P. J., Fryer, T. D., Smielewski, P., and Hodges, J. R. (2003). Limbic hypometabolism in Alzheimer's disease and mild cognitive impairment. *Ann. Neurol.* **54**(3), 343–351.

Neuropathology Group of the Medical Research Council Cognitive Function and Ageing Study (MRC-CFAS) (2007). Pathological correlates of late-onset dementia in a multicentre, community-based population in England and Wales. *Lancet* **357**(9251), 169–175.

Ng, S., Villemagne, V. L., Berlangieri, S., Lee, S. T., Cherk, M., Gong, S. J., Ackermann, U., Saunder, T., Tochon-Danguy, H., Jones, G., Smith, C., O'Keefe, G., *et al.* (2007). Visual assessment versus quantitative assessment of 11C-PIB PET and 18F-FDG PET for detection of Alzheimer's disease. *J. Nucl. Med.* **48**(4), 547–552.

Pathological correlates of late-onset dementia in a multicentre, community-based population in England and Wales (2001). Neuropathology Group of the Medical Research Council Cognitive Function and Ageing Study (MRC CFAS). *Lancet* **357**(9251), 169–75.

Patwardhan, M. B., McCrorg, D. C., Matchar, D. B., Samsa, G. P., and Rutschmann, O. T. (2004). Alzheimer disease: Operating characteristics of PET—A meta-analysis. *Radiology* **231**(1), 73–80.

Petersen, R. C., Smith, G. E., Waring, S. C., Ivnik, R. J., Tangalos, E. G., and Kokmen, E. (1999). Mild cognitive impairment: Clinical characterization and outcome. *Arch. Neurol.* **56**(3), 303–308.

Petersen, R. C., Stevens, J. C., Ganguli, M., Tangalos, E. G., Cummings, J. L., and DeKosky, S. T. (2001). Practice parameter: Early detection of dementia: Mild cognitive impairment (an evidence-based review). Report of the Quality Standards Subcommittee of the American Academy of Neurology. *Neurology* **56**(9), 1133–1142.

Phelps, M. E., Mazziotta, J. C., and Huang, S. C. (1982). Study of cerebral function with positron computed tomography. *J. Cereb. Blood Flow Metab.* **2**(2), 113–162.

Pike, K. E., Savage, G., Villemagne, V. L., Ng, S., Moss, S. A., Maruff, P., Mathis, C. A., Klunk, W. E., Masters, C. L., and Rowe, C. C. (2007). Beta-amyloid imaging and memory in non-demented individuals: Evidence for preclinical Alzheimer's disease. *Brain* **130**(Pt 11), 2837–2844.56.

Pimlott, N. J., Siegel, K., Persaud, M., Slaughter, S., Cohen, C., Hollingworth, G., Cummings, S., Drummond, N., Dalziel, W., Sylvius, J., Pringle, D., and Eliasziw, T. (2006). Management of dementia by family physicians in academic settings. *Can. Fam. Physician* **52**(9), 1108–1109.

Riedel-Heller, S. G., Schork, A., Matschinger, H., and Angermeyer, M. C. (1999). The role of referrals in diagnosing dementia at the primary care level. *Int. Psychogeriatr.* **11**(3), 251–262.

Silverman, D. H., Gambhir, S. S., Huang, H. W., Schwimmer, J., Kim, S., Small, G. W., Chodosh, J., Czernin, J., and Phelps, M. E. (2002). Evaluating early dementia with and without assessment of regional cerebral metabolism by PET: A comparison of predicted costs and benefits. *J. Nucl. Med.* **43**(2), 253–266.

Silverman, D. H., Small, G. W., Chang, C. Y., Lu, C. S., Kung De Aburto, M. A., Chen, W., Czernin, J., Rapoport, S. I., Pietrini, P., Alexander, G. E., Schapiro, M. B., Jagust, W. J., *et al.*

(2001). Positron emission tomography in evaluation of dementia: Regional brain metabolism and long-term outcome. *JAMA* **286**(17), 2120–2127.
Tang, M. X., Stern, Y., Marder, K., Bell, K., Gurland, B., Lantigua, R., Andrews, H., Feng, L., Tycko, B., and Mayeux, R. (1998). The APOE-epsilon4 allele and the risk of Alzheimer disease among African Americans, whites, and Hispanics. *JAMA* **279**(10), 751–755.
UN (2003). "The United Nations Population Division. World Population Prospects: The 2002 Revision." United Nations, New York.
US Preventive Services Task Force (2003). Screening for dementia: Recommendations and rationale. Guide to Clinical Prevention Services. *Ann. Intern. Med.* **139,** 925–926.
Wimo, A., Winblad, A., and Jönsson, A. (2007). An estimate of the total worldwide societal costs of dementia in 2005. *Alzheimers Dement. J. Alzheimers Assoc.* **3**(2), 81–91.
World Health Organization (2005). "Mental Health Atlas." p. 471.

THE MOLECULAR AND CELLULAR PATHOGENESIS OF DEMENTIA OF THE ALZHEIMER'S TYPE: AN OVERVIEW

Francisco A. Luque*,† and Stephen L. Jaffe*

*Department of Neurology, Louisiana State University School of Medicine-Shreveport, Shreveport, Louisiana 71103, USA

†VA Neurology Service, Overton Brooks VAMC, Shreveport, Louisiana 71101, USA

The pathogenesis of dementia of the Alzheimer's type (DAT) remains elusive. The neurodegeneration occurring in this disease has been traditionally believed to be the result of toxicity caused by the accumulation of insoluble amyloid-beta 42 (AB) aggregates, however recent research questions this thesis and has suggested other more convincing cellular and molecular mechanisms. Dysfunction of amyloid precursor protein metabolism, AB generation/aggregation and/or degredation/clearance, tau metabolism, protein trafficking, signal transduction, heavy metal homeostasis, acetylcholine and cholesterol metabolism, have all been implicated etiologically especially as to production of neurotoxic by-products occurring as a result of a specific process derangement. In this paper, these and other research directions are discussed as well as their implications for future therapies. The relationship of the proposed abnormal molecular and cellular processes to underlying genetic mutations is also scrutinized, all in an attempt to stimulate further insight into the pathogenesis of, and thus therapeutics for this increasingly prevalent disease.

I. Introduction

Over 100 years ago, Alzheimer described a patient with memory disturbance and a brain pathological picture which included miliary bodies (plaques) and dense bundles of fibrils (tangles). These observations remain the clinical and pathological hallmarks of dementia of the Alzheimer's type (DAT). DAT is

INTERNATIONAL REVIEW OF NEUROBIOLOGY, VOL. 84
DOI: 10.1016/S0074-7742(09)00408-5

the most common form of dementia and accounts for 50-60% of dementia presentations. Presently there are approximately 5.1 million patients in the USA with DAT; and it is estimated that in the next 20 years, there will be an additional 15 million cases. DAT incidence increases with age. Before age 60, the incidence is about 5%; but after age 85, it increases to 50%. The annual cost attributed to DAT in this country is $148 billion. The clinical symptoms of DAT include alterations of abstract thinking, skilled movements, emotional expression, executive function, and memory. The majority of DAT cases are sporadic, but around 10% are inherited. Autosomal dominant cases are related to mutations in the amyloid precursor protein (APP) (chromosome 21), presenilin 1 (PS1) (chromosome 14), and presenilin 2 (PS2) (chromosome 1) genes. Similar pathology with a dementing process often occurs in Down syndrome (trisomy 21).

The pathogenesis of DAT remains perplexing. Dysfunction in APP metabolism, $A\beta 42$ ($A\beta$) degradation and clearance, signal transduction, tau metabolism, protein trafficking, acetylcholine and cholesterol metabolism, and homeostasis of heavy metals may be involved (Bertram and Tanzi, 2008). DAT neurodegeneration may be linked to dysfunction of the degradation process of $A\beta$ or other proteins, or $A\beta$ generation with production of toxic by-products may be responsible, and aggregate formation may be a benign phenomenon. On the other hand, $A\beta$ aggregates may be primarily toxic interfering with the degradation of various proteins by interfering with chaperone function. Protein folding, protein-protein interaction, and protein membrane association are all thermodynamically driven processes utilizing enzymes known as chaperones. The importance of chaperones for proteosomal degradation has been corroborated in the endoplasmic reticulum (ER) where newly synthesized proteins that are abnormal or misfolded are removed by a quality control mechanism termed ER-associated degradation (Tai and Schuman, 2008; Zhang *et al.*, 2004).

II. Cellular Pathogenesis and Investigational Strategies

DAT histological brain changes are especially prevalent in the frontal and temporal lobes and include neuronal loss, extracellular deposition of $A\beta$ as "senile plaques," and the intracellular accumulation of neurofibrillary tangles composed of microtubule-associated protein tau arrayed as paired helical filaments. One of the earliest neuropathological changes is the presence of a large number of reactive astrocytes at the site of $A\beta$ deposition. $A\beta$ deposition also leads to degeneration of the microvascular basement membrane and thus alteration in blood-brain barrier permeability (Berzin *et al.*, 2000) (Fig. 1).

As previously noted, it is unclear whether the process of formation and accumulation, precursors occurring during that process, the actual accumulated $A\beta$, and/or abnormal degradation initiates a series of neurotoxic events including

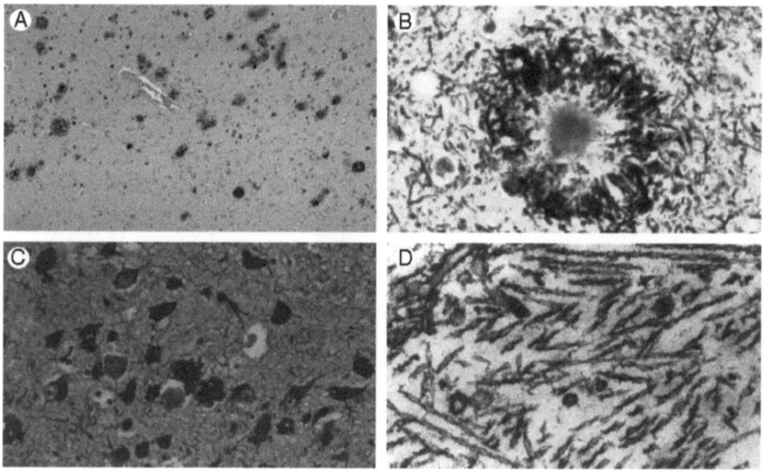

FIG. 1. Neuropathology of Alzheimer's disease. (A) Low power amyloid plaques, (B) high power amyloid plaque, (C) neurofibrillary tangles, silver stained, and (D) electron micrograph of neurofibrillary tangles composed of hyperphosphorylated tau (Courtesy of Sisodia and St. George-Hyslop, 2002; with permission from Nature Publishing Group). (See Color Insert.)

hyperphosphorylation of tau that triggers neuronal dysfunction and cell apoptosis (the "amyloid cascade"). Although new ligands for positron emission tomographic (PET) imaging of Aβ *in vivo* (Pittsburgh Compound-B) are available, their diagnostic value is questionable due to this conundrum. However, the accumulation of amyloid plaques is the earliest feature that can be detected in presymptomatic individuals with a PS1 mutation or with trisomy 21. Although, it is not clear how Aβ specifically induces neurodegeneration, the accumulation of intracellular tau does appear to be neurotoxic.

It was originally thought that plaque formation occurred over long periods of time and long before the dementia became apparent. However, Meyer-Luehmann *et al.* (2008) using a DAT mouse model provided evidence that amyloid plaques form extremely rapidly, actually within 24 h, and their size and final characteristics stabilize within a week. In the early stages of the disease, microplaques can damage neighboring axons and dendrites within days. Among the features of the DAT degenerative process are damage to synaptic connections associated with accumulation of amyloid β oligomers, and neuritic dystrophy—the formation of tortuous neuronal elongations with eventual loss of selective groups of neurons. These processes have been strongly linked to DAT cognitive impairment.

Defects in axonal transport have been described in animal models of DAT including transgenic mice overexpressing APP, and those containing mutant PS1. However, the mechanism is not fully understood (De Vos *et al.*, 2008). Aβ appears to disrupt the axonal transport of a variety of cargoes including mitochondria by associating with and thus damaging the mitochondria (Hiruma *et al.*, 2003; Manczak *et al.*, 2006). Mutant PS1 impairs axonal transport by causing a reduction in kinesin-I driven motility via interaction with glycogen synthase kinase-3β to produce phosphorylation of kinesin light chains with release of kinesin-I from membrane-bound organelles (Pigino *et al.*, 2003).

Disturbance of glutaminergic neurotransmission and consequent excitotoxicity has been implicated in the progression of DAT (Matos *et al.*, 2008). In DAT patients, increased glutaminergic excitotoxic activity appears to occur due to Aβ interference with glutamate uptake by astrocytes, leading to increased activation of NMDA neuronal receptors. Memantine is a noncompetitive NMDA receptor antagonist that appears to protect neurons from glutamate mediated excitotoxicity without interfering with the physiological NMDA receptor activation needed for cognitive functioning. Although there is no data showing memantine has a beneficial effect in mild DAT, this drug is well tolerated and may be a useful adjunct in moderate to severe disease defined as a mini-mental status exam (MMSE) score of less than 15.

Amyloid β aggregation disturbs the cellular proteosome by inducing oxidative stress with subsequent mitochondrial and proteosomal dysfunction (Hamacher *et al.*, 2007). Mitochondrial dysfunction appears to be an early causative event in neurodegeneration. Mutations of mitochondrial fusion GTPases, mitofusin 2, and optic atrophy 1, are neurotoxic and produce oxidative stress which disrupts the cable-like morphology of functional mitochondria. This causes impairment of bioenergetics, interfering with mitochondrial migration and triggering neurodegeneration (Knott *et al.*, 2008).

Sortilin related receptor (SorLA) belongs to a family of sorting receptors that mediate various intracellular processing and trafficking functions (Yamazaki *et al.*, 1996). SorLA/LR11 mediates re-internalization of APP, during which APP is transported from the cell surface to an endocytic compartment where it is cleaved into Aβ by β secretase and γ secretase. SorLA/LR11 is highly expressed in the brain and binds to ApoE and APP, sequestering APP in the Golgi apparatus and thus preventing APP generation into Aβ (Andersen *et al.*, 2006). In all studies to date, there is no consistent evidence of a SorLA association with DAT, possibly due to allelic heterogeneity. Studies *in vitro* suggest that SorLA expression is reduced in the brains of DAT patients, even in the preclinical state. SorL1-knockout mice show increase in brain Aβ level. Although there is convergence of data from genetic and biochemical studies, more data are needed to better evaluate this association.

Transferrin (TF) is a major circulating glycoprotein involved in the metabolism of iron and is highly expressed in the brain. Iron misregulation promotes neurodegeneration via generation of reactive oxygen species (ROS) (Brewer, 2007). Iron is increased in the brains of DAT patients where it is associated with plaques and tangles. Iron may play a role in the aggregation of hyperphosphorylated tau into insoluble paired helical filaments (Yamamoto *et al.*, 2002), and iron may also regulate the translation of APP. Thus iron levels could modulate the generation of both plaques and tangles. Besides the binding of metals to Aβ, TF has been shown to modulate several physiochemical properties of Aβ including the rate of aggregation.

The ability of neurons to regulate the influx, efflux, and subcellular compartmentalization of calcium is compromised in DAT. This appears due to age-related oxidative stress, and impaired metabolism/accumulation of Aβ oligomers. Aβ can promote cellular calcium overload via oxidative stress and by forming pores in the membrane, rendering neurons vulnerable to excitotoxicity and apoptosis (Bezprozvanny *et al.*, 2008). Mutations of presenilin genes that cause early onset DAT also cause ER (endoplasmic reticulum) calcium overload by impairing the normal ER calcium leak-channel function of the presenilins. A perturbation in calcium homeostasis may be involved in early DAT by altering APP processing and thus Aβ production. Calcium activated neural proteinases (CANPs or calpains) are key enzymes in the intracellular signaling cascade and thus may mediate calcium dependent neural degeneration. Pharmacological modulation of the CANP system could be a potential therapeutic strategy in Alzheimer's disease (Saito *et al.*, 1993).

De Luigi *et al.* (2002) studied the presence of circulating cytokines and the ability of blood cells to release them in response to inflammatory stimuli in patients with different forms of dementia. A significant increase in circulating interleukin-1β was found in moderate DAT. Chronic inflammation as found in head trauma is an important risk factor for DAT, and the inflammatory process affects glial-neuronal interactions. The activation of the glial inflammatory process, either caused by genetic or environmental insults to neurons, produces glial derived cytokines and other proteins that results in neurodegeneration. Interleukin 1 is key in initiating and coordinating neuronal synthesis and in the processing of APP, resulting in the continuous deposition of Aβ. It also promotes astrocytic synthesis and release of inflammatory and neuroactive molecules such as S100β, which is a neurite growth promoting cytokine. This cytokine can stress neurons fostering neuronal cell dysfunction, causing apoptosis by increasing intraneuronal free calcium concentrations. This neuronal injury effect activates microglial cells which then overexpress inteleukin 1, thereby producing feedback amplification and self-propagation of this cytokine cycle (Griffin *et al.*, 1998).

III. Molecular and Genetic Studies Resulting in Therapeutic Strategies

Amyloid β is generated by endoproteolysis of the APP. This is achieved by cleavage of APP by a group of enzyme complexes: α, β, and γ secretases. Three enzymes with α secretase activity belong to the ADAM enzyme family (desintegrin and methaloproteinase enzymes): ADAM9, ADAM10, and ADAM17 (i.e., tumor necrosis factor converting enzyme). The β site APP cleaving enzyme-1 (BACE1), β secretase, is a type I integral membrane protein from the pepsin family of aspartyl proteases. Gamma secretase is an intramembranous complex of enzymes composed of presenilin 1 or 2, nicastrin, anterior pharynx defective-1, and presenilin enhancer 2, with the presenilin being the active site (Wolfe *et al.*, 1999; Yu *et al.*, 2000).

There are two pathways in the cleavage and processing of APP. One is non-amyloidogenic. APP is cleaved by α secretase at amino acid position 83, the carboxy terminus, producing a large amino terminal ectodomain (sAPPα). This is secreted into the extracellular medium where it is in turn cleaved by γ secretase producing a short peptide, p3. The cleavage by α secretase is in the middle of the APP region, and therefore there is no Aβ production. The amyloidogenic pathway leads to Aβ generation. Initially the β secretase cleaves at amino acid position 99, releasing sAPP β into the extracellular space. Meanwhile the C99 fragment stays in the membrane where it is cleaved by γ secretase between residues 38 and 43, liberating a peptide which is 40 amino acids in length (Aβ40). Only a small proportion (10%) composes the 42 residue, the Aβ peptide (Aβ42). Aβ is more hydrophobic and thus prone to form fibrils, and is abundant in senile plaques (Jarrett *et al.*, 1993; Martin, 1999; Suh and Checler, 2002; Younkin, 1998). Mutations in the APP gene linked to early-onset DAT produce a product that is preferentially cleaved to release Aβ 42 (Selkoe, 2001; Sisodia and St. George-Hyslop, 2002) (see Fig. 2).

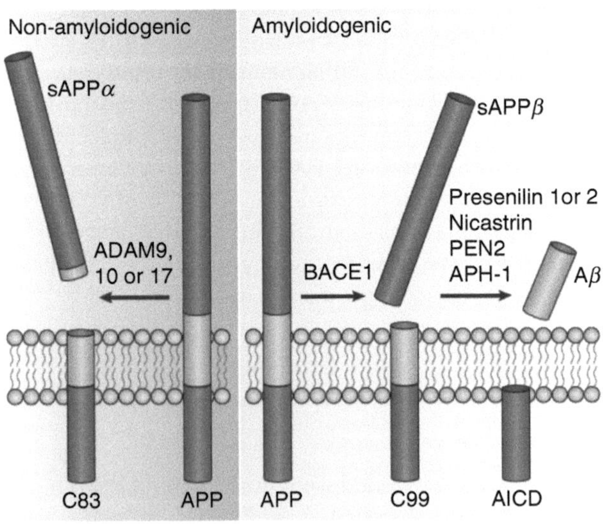

FIG. 2. Pathways in the cleavage and processing of APP (Courtesy of LaFerla *et al.*, 2007; with permission from Nature Publishing Group).

Presently, the major investigational therapeutic strategies are focused on inhibiting brain Aβ production, aggregation, and/or increasing Aβ brain clearance. Secretase modulators are being studied as Aβ production inhibitors based on the observation that Aβ production is abolished in BACE1 knockout mice and the DAT phenotype does not develop. BACE1 inhibitors have been developed to reduce brain Aβ accumulation in DAT transgenic mice. Rajendra *et al.* (2008) focused not only on inhibition of β secretase active site binding, but also on inhibiting the localized active enzyme at the subcellular level. He synthesized a membrane-anchored version of β secretase transition-state inhibitor by linking it to a sterol moiety. This targeted the inhibition of active β secretase found in endosomes, reducing the enzyme activity more readily than free inhibitor alone. Similarly γ secretase inhibitors specific to the APP cleavage site have been developed because of the concern of blocking γ secretase activity on other substrates such as the notch protein. Testing in transgenic mice is underway and appears successful in blocking the DAT phenotype, while inhibitors of Aβ accumulation show no effect on phenotype development again suggesting that effects of the process or the actual precursors may be the etiologic factors of DAT neurotoxicity (Schilling *et al.*, 2008).

Drugs that stimulate α secretase thus shifting the APP cleavage to a non-amyloidogenic pathway also reduce Aβ production. Among these, Bryostatin, a protein kinase C activator used for cancer treatment, increases α secretase activity and reduces Aβ accumulation in DAT transgenic mice models. Removal of Aβ by immunization is based on the observation that active immunization with fibrillar Aβ reduces Aβ deposition in DAT transgenic mice. In similar fashion, using a passive antibody technique against Aβ also reduces Aβ deposition. It is postulated that Aβ antibodies bind to Aβ plaques and induce Aβ clearance by microglia. However, a vaccine trial in humans was halted when 6% of patients developed encephalitis apparently due to T cell response to the C-terminus of the vaccine peptide. Presently monoclonal antibodies are being studied as well as nanobodies in attempts to increase Aβ clearance.

Inhibitors of Aβ fibrillization are among peptides that interfere with the assembly of Aβ into oligomers which eventually form protofibrils and fibrils. Two peptides have been shown to prevent the interaction Aβ-Aβ and Aβ-ApoE without inducing an immune response (Permanne *et al.*, 2002; Sadowski *et al.*, 2004). Glycosaminoglycans bind Aβ and can promote aggregation of Aβ. The drug NC-531 (Tramiprosate or Alzhemed) is a glycosaminoglycan mimetic designed to do the opposite and block Aβ aggregation. However, phase III clinical trials did not demonstrate a therapeutic effect for this compound. Scyllo-cyclohexanediol inhibits Aβ 42 fibril assembly, and tetracyclines as well as rifampin inhibit amyloid formation *in vitro*. Terenflurbil (Flurizan) is an allosteric modulator of γ secretase, inhibiting its activity. It is a stereoisomer of a nonsteroidal anti-inflammatory that does not have cyclo-oxygenase (COX) activity. Again however, a phase III trial failed to show clinical benefit.

Copper and Zinc ions can induce Aβ aggregation. The chelator, clioquinol (PBT-1), reduces brain deposition of Aβ in DAT transgenic mice. Clinical trials were stopped when compound impurities produced toxicity. However, PBT-2 (a second generation metal protein attenuating compound which does not contain iodine) is now undergoing clinical trial. In a DAT transgenic mouse model, preliminary results suggested improvement in a subset of cognitive measures, as well as reduction in cerebrospinal fluid and brain Aβ. Apomorphine and its derivatives promote the oligomerization of Aβ but inhibit the fibrillization. Structural activity demonstrated that 10, 11-dihydroxy substitutions in the D ring of apomorphine are required for the inhibitory effects. Methylation of these hydroxyl groups reduces the inhibitory potency. The ability of these molecules to inhibit Aβ formation appears to be linked to their tendency to undergo rapid auto-oxidation, suggesting that auto-oxidation products act directly or indirectly on Aβ and inhibit fibrillization. This paradigm offers the potential for designing more efficient inhibitors of Aβ amyloidogenesis *in vivo* (Lashuel *et al.*, 2002).

Nonsteroidal anti-inflammatory drugs (NSAIDs) reduce brain Aβ accumulation in DAT transgenic mice. This effect may be due to inhibition of COX or via direct effect on γ secretase. However, clinical trials with typical NSAIDs and COX-2 inhibitors (celocoxib and refecoxib) as well as other anti-inflammatories including prednisone and hydroxychloroquine, in established DAT showed no positive effects on cognition levels. It is postulated that this failure was due to the lack of ability to reverse already established pathology, and that the intervention has to be in the early stages of mild cognitive impairment (MCI). Statins also reduce brain Aβ accumulation in DAT transgenic mice. However, prospective studies showed no association between the use of statins and the risk of DAT development, nor were there any changes in CSF Aβ levels. Large, extended, randomized clinical trials are needed to determine if any benefit can be obtained with this drug class.

Nitric oxide may be involved in DAT neurodegeneration. By using redox proteomic techniques, 10 proteins have been associated with an increase of nitrotyrosine immunoreactivity in the brains of DAT patients. Of these proteins, α-enolase has been shown to be oxidized in the brains of DAT patients (Calabrese *et al.*, 2007; Castegna *et al.*, 2002). It is one of the subunits of enolase which catalizes the reversible conversion of 2-phosphoglycerate to phosphoenolpyruvate in glycolysis. This, in addition to the increase of nitration of the triosephosphate isomerase that interconverts dihydroxyacetone phosphate and 3-phosphoglyceraldehyde in glycolysis, may explain the altered glucose tolerance and metabolism seen in DAT patients. Neuropeptide h3 (a phosphatidylethanolamine-binding protein), hippocampal cholinergic neurostimulating peptide, and raf-kinase inhibitor protein have various functions in the brain. *In vitro*, they upregulate levels of choline acetyltransferase in cholinergic neurons following *N*-methyl D-aspartate (NMDA) receptor activation. The activity of this enzyme is reduced in DAT patients. Together with cholinergic deficits, the nitration of neuropolypeptide h3 and the subsequent loss of neurotrophic action on cholinergic neurons of the hippocampus and basal forebrain may be part of the explanation of cognitive decline in DAT (Giacobini, 2003; Rossor *et al.*, 1982).

The polyphenolic molecule, curcumin, shows promise in reducing nitrosative brain injury and delaying the onset of neurodegenerative disorders. It is a strong antioxidant that inhibits lipid peroxidation intercepting and neutralizing ROS and reactive nitrogen species (RNS) and increasing haeme-oxygenase I expression in astrocytes and neurons. Dietary curcumin suppressed indicators of inflammation and oxidative damage in the brains of DAT transgenic mice (Butterfield *et al.*, 2002; Scapagnini *et al.*, 2006). However, antioxidants including vitamin E did not show any effect on the rate of progression of MCI to DAT in large studies. Drugs that reduce tau phosphorylation by inhibiting tau such as CDK5 (cyclin-dependent kinase 5) and GSK-3β (glycogen synthase kinase 3β) are being tested. Since tau phosphorylation results from the action of multiple kinases and phosphatases, these drugs may not effect normalization of tau phosphorylation.

Despite the identification of mutations in three genes associated with early-onset DAT (i.e., the APP, PS1, and PS2 genes) and one genetic risk factor for sporadic DAT (the gene for apolipoprotein E), our understanding of the pathological genetic mechanism remains elusive. There are additionally over 500 different gene candidates which may be involved in DAT pathogenesis, and at least 20 loci with modest but significant effects on the risk for this disease (Bertram and Tanzi, 2008). The APP gene is located at chromosome 21q21.3. Twenty-nine mutations in 78 affected families have been identified. These mutations are close to the β and γ secretase APP cleavage sites (Goate *et al.*, 1991). The PS1 gene is located at chromosome 14q24.3 and 166 mutations have been identified in 362 families. These mutations are associated with an increase in the Aβ42/Aβ40 ratio, and are located throughout the gene for the PS1 molecule where γ secretase complexing occurs (Sherrington *et al.*, 1995). PS2 localizes to chromosome 1q31-42. Ten mutations have been identified in 18 affected families. There is an increase in the Aβ42/Aβ40 ratio, similar to the effects of PS1 gene mutations. The mutations are found throughout the gene, and thus γ secretase complexing is affected (Levy-Lahad *et al.*, 1995; Rogaev *et al.*, 1995).

Roberson *et al.* (2007) studied a hybrid mouse expressing mutant APP and having the tau gene inactivated. The elimination of tau did not alter the accumulation of senile plaques but did prevent the memory loss, behavioral disturbances and sudden death associated with this DAT-like phenotype. The complete elimination of tau was not necessary, a reduction of 50% producing significant improvements. On the other hand, soluble Aβ assemblies cause memory loss and bind to dendritic spines. It would appear that soluble Aβ assemblies and not the precipitated senile plaques are the cause of impaired cognition, but the exact mechanism remains ill-defined (Lacor *et al.*, 2004). Roberson appropriately sounded a note of caution in extrapolating mouse data into human pathophysiology in which other compensatory mechanisms may be playing a role (Avila *et al.*, 2004).

Bertram and Tanzi (2008) discussed 10 different selected genes potentially implicated in DAT. Angiotensin converting enzyme (ACE) has been shown *in vitro* to degrade naturally secreted Aβ and this could explain the increased risk of DAT in carriers of the ACE I allele; however, the relevance *in vivo* needs to be demonstrated. CH25H (cholesterol 25-hydrolase) which is responsible for the synthesis of 25-hydroxycholesterol, is a potent regulator of transcription of genes involved in cholesterol and lipid metabolism. There seems to be an association between variants of CH25H and DAT. One study (Papassotiropoulus *et al.*, 2005) suggested an increased DAT risk associated with a certain CH25H haplotype (containing the rs13500 risk allele) leading to an increase in CSF lathosterol (a metabolite precursor of cholesterol) as well as a higher brain Aβ load with lower levels of Aβ in the CSF of non-demented elderly subjects. Both CH25H and lipase A (LIPA) as well as APOE are involved in cholesterol metabolism, and Aβ production and clearance have been shown to be regulated by cholesterol. Drugs that inhibit cholesterol synthesis lower Aβ in cellular and animal models. Nicotinic acetylcholine receptors are widely expressed in the brain, and the β2 (CHRNB2) subunit is particularly abundant and forms a heteropentameric receptor ($\alpha 4\beta 2$) with the α4 subunit. The CHRNB2 gene which encodes the nAChR β2 subunit maps to chromosome 1q21. Two independent genetic association studies have investigated the potential involvement of this gene in DAT risk. Wu *et al.* (2004) found that Aβ directly inhibits the $\alpha 4\beta 2$ receptor complex, and that nicotine possibly by activating the NACHR can mitigate the toxic effects of Aβ, as well as influencing tau phosphorylation *in vitro* and *in vivo* (Oddo *et al.*, 2005).

Grb-2-associated binder 2 (GAB2) is a highly conserved scaffolding/adaptor protein involved in signaling pathways and in particular in the transduction of cytokine and growth-receptor signaling (Liu and Rohrschneider, 2002; Sarmay *et al.*, 2006). GAB2 is expressed ubiquitously, although in greatest amounts in white blood cells, the prefrontal cortex, and the hypothalamus. All 10 GAB2 gene single nucleotide polymorphisms (SNPs) showed significant association with DAT. A protective role is predicated for all the minor alleles. All 10 of the SNPs display a high LOD score ($\log_{10}$ ratio of probability data occurring if loci linked, to probability data occurring with unlinked loci), and therefore probably point to a single underlying signal effect. Of the 10 associated markers, only one, rsl1385600, is predicted to map within the coding region for GAB2 where it does not appear to produce a change in amino acid sequence. It has been suggested that changes in GAB2 expression could potentially affect glycogen synthase kinase 3 (GSK3)-dependent phosphorylation of tau, and thus formation of neurofibrillary tangles. Moreover, growth factor receptor-bound protein 2 which binds GAB2 has been reported to bind tau, APP and PS1 and 2 (Nizzari *et al.*, 2007).

Laminins are intermediate filament proteins that are a component of the nuclear lamina. Alternative splicing produces two different isoforms, laminin A (LMNA) and C. Mutations in LMNA have been found to cause a number of different disorders. The LMNA gene maps 1.5 Mb distal to the CHRNB2 gene on chromosome 1, and close to the presumed DAT linkage region at 1q22-q25. Only one group has reported an association between variants of LMNA and DAT (Grupe *et al.*, 2007), the LMNA association ranked second in terms of genetic effect size of all non-APOE-4 related genes with an OR (odds ratio) of 1.35.

Macrotubule-associated protein tau (MAPT) is found in NFTs. The gene that encodes MAPT has been suspected of harboring disease causing mutations, but the search for such mutations in DAT has been unsuccessful. However, MAPT mutations were found to cause another form of dementia, frontotemporal lobar degeneration with parkinsonism linked to chromosome 17 (FTDP-17). Despite the lack of MAPT mutation associations in DAT, it is likely that tau dysfunction still contributes to an increased risk of developing DAT (Ballatore *et al.*, 2007).

Prion protein (PRP) is a membrane tethered glycoprotein that is highly expressed in all regions of the brain, mostly in neurons. Its physiological functions are not clear. Mutations of prion protein gene (PRNP) are a major determinant of familial and sporadic prion diseases such as Creutzfeldt-Jakob disease. Most prion diseases produce a rapidly progressive neurodenegeneration with spongioform brain changes and amyloid plaques that consist of misfolded PRP aggregates. There are more than two dozen different amino acid substitution mutations identified, which are transmitted in an autosomal dominant fashion with 100% penetrance. Aβ positive plaques in DAT brains often contain PRP deposits. Recent evidence from studies using APP-PRNP double transgenic mice suggests that PRP might actually promote plaque formation in DAT by increasing Aβ aggregation (Schwarze-Eicker *et al.*, 2005).

References

Andersen, O. M., Schmidt, V., Spoelgen, R., Gliemann, J., Behlke, J., Galatis, D., Mckinstry, J., Parker, M. W., Masters, C. L., Hyman, B. T., Cappai, R., and Willnow, T. E. (2006). Molecular dissection of the interaction between amyloid precursor protein and its neuronal trafficking receptor Sort LA/LR11. *Biochemistry* **45,** 2618–2628.

Avila, J., Lucas, J. J., Perez, M., and Hernandez, F. (2004). Role of tau protein in both physiological and pathological conditions. *Physiol. Rev.* **84,** 361–384.

Ballatore, C., Lee, V. M., and Trojanowski, J. Q. (2007). Tau mediated neurodegeneration in Alzheimer's disease and related disorders. *Nat. Rev. Neurosci.* **8,** 663–672.

Bertram, L., and Tanzi, R. E. (2008). Thirty years of Alzheimer's disease genetics: The implications of systematic meta-analyses. *Nat. Rev. Neurosci.* **9,** 768–778.

Berzin, T. M., Zipser, B. D., Rafii, M. S., Kuo-Leblanc, V., Yancopoulos, G. D., Glass, D. J., Fallon, J. R., and Stopa, E. G. (2000). Agrin and microvascular damage in Alzheimer's disease. *Neurobiol. Aging* **21,** 349–355.

Bezprozvanny, I., and Mattson, M. P. (2008). Neuronal calcium mishandling and the pathogenesis of Alzheimer's Disease. *Trends Neurosci.* **31,** 454–463.

Brewer, G. J. (2007). Iron and copper toxicity in diseases of aging, particularly atherosclerosis and Alzheimer's disease. *Exp. Biol. Med. (Maywood)* **232,** 323–335.

Butterfield, D. A., Castegna, A., Pocernich, C. B., Drake, J., Scapagnini, G., and Calabrese, V. (2002). Nutritional approaches to combat oxidative stress in Alzheimer's disease. *J. Nutr. Biochem.* **13,** 444.

Calabrese, V., Mancuso, C., Calvani, M., Rizzarelli, E., Butterfield, D. A., and Stella, A. M. (2007). Nitric oxide in the central nervous system: Neuroprotection versus neurotoxicity. *Nat. Rev. Neurosci.* **8,** 766–775.

Castegna, A., Aksenov, M., Thongboonkerd, V. R., Klein, J. B., Pierce, W. M., Booze, R., Markesbery, W. R., and Butterfield, D. A. (2002). Proteomic identification of oxydatively modified proteins in Alzheimer's disease brain. Part II: Dihydropyrimidinase related protein 2, alpha-enolase and heat shock cognate 71. *J. Neurochem.* **82,** 1524–1532.

De Luigi, A., Pizzimenti, S., Quadri, P., Lucca, U., Tettamanti, M., Fragiacomo, C., and De Simoni, M. G. (2002). Peripheral inflammatory response in Alzheimer's disease and multi-infarct dementia. *Neurobiol. Dis.* **11,** 308–314.

De Vos, K. J., Grierson, A. J., Ackerley, S., and Miller, C. C. (2008). Role of axonal transport in neurodegenerative diseases. *Annu. Rev. Neurosci.* **31,** 151–173.

Giacobini, E. (2003). Cholinergic function and Alzheimer's disease. *Int. J. Geriat. Psychiatry* **18,** S1–S5.

Goate, A., Chartier-Harlin, M. C., Mullan, M., Brown, J., Crawford, F., Fidani, L., Gluffra, L., Haynes, A., Irving, N., James, L., Mant, R., Newton, P., *et al.* (1991). Segregation of a missense mutation in the amyloid precursor protein gene with familial Alzheimer's disease. *Nature* **349,** 704–706.

Griffin, W. S. T., Sheng, J. G., Royston, M. C., Gentleman, S. M., Mckenzie, J. E., Graham, D. I., Roberts, G. W., and Mrak, R. E. (1998). Glial-neuronal interactions in Alzheimer's disease: The potential role of a 'cytokine cycle' in disease progression. *Brain Pathol.* **8,** 65–72.

Grupe, A., Abraham, R., Li, Y., Rowland, C., Hollingworth, P., Morgan, A., Jehu, L., Segurado, R., Stone, D., Schadt, E., Karnoub, M., Nowotny, P., *et al.* (2007). Evidence for novel susceptibility genes for late-onset Alzheimer's disease from a genome-wide association study of putative functional variants. *Hum. Mol. Genet.* **16,** 865–873.

Hamacher, M., Meyer, H. E., and Marcus, K. (2007). New access to Alzheimer's and other neurodegenerative diseases. *Expert Rev. Proteomics* **4,** 591–594.

Hiruma, H., Katakura, T., Takahashi, S., Ichikawa, T., and Kawakami, T. (2003). Glutamate and amyloid beta-protein rapidly inhibit fast axonal transport in cultured rat hippocampal neurons by different mechanisms. *J. Neurosci.* **23,** 8967–8977.

Jarrett, J. T., Berger, E. P., and Lansbury, P. T., Jr. (1993). The carboxy terminus of the β amyloid protein is critical for seeding of amyloid formation: Implications for the pathogenesis of Alzheimer's disease. *Biochemistry* **32,** 4693–4697.

Knott, A. B., Perkins, G., Schwarzenbacher, R., and Bossy-Wetzel, E. (2008). Mitochondrial fragmentation in neurodegeneration. *Nat. Rev. Neurosci.* **9,** 505–518.

Lacor, P. N., Buniel, M. C., Chang, L., Fernandez, S. J., Gong, Y., Viola, K. L., Lambert, M. P., Velasco, P. T., Bigio, E. H., Finch, C. E., Kraft, G. A., and Klein, W. L. (2004). Synaptic targeting by Alzheimer's-related amyloid beta oligomers. *J. Neurosci.* **24,** 10191–10200.

LaFerla, F. M., Green, K. N., and Oddo, S. (2007). Intracellular amyloid-beta in Alzheimer's disease. *Nat. Rev. Neurosci.* **8,** 499–509.

Lashuel, H. A., Hartley, D. M., Balakhaneh, D., Aggarwal, A., Teichberg, S., and Callaway, D. J. (2002). New class of inhibitors of amyloid-beta fibril formation. Implications for the mechanism of pathogenesis in Alzheimer's disease. *J. Biol. Chem.* **277,** 42881–42890.

Levy-Lahad, E., Wasco, W., Poorkaj, P., Romano, D. M., Oshima, J., Pettingell, W. H., Yu, C. E., Jondro, P. D., Schmidt, S. D., and Wang, K. (1995). Candidate gene for the chromosome 1 familial Alzheimer's disease locus. *Science* **269,** 973–977.

Liu, Y., and Rohrschneider, L. R. (2002). The gift of Gab. *FEBS Lett.* **515,** 1–7.

Manczak, M., Anekonda, T. S., Henson, E., Park, B. S., Quinn, J., and Reddy, P. H. (2006). Mitochondria are a direct site of A beta accumulation in Alzheimer's disease neurons: Implications for free radical generation and oxidative damage in disease progression. *Hum. Mol. Genet.* **15,** 1437–1449.

Martin, J. B. (1999). Molecular basis of the neurodegenerative disorders. *N. Engl. J. Med.* **340,** 1970–1980.

Matos, M., Augusto, E., Oliveira, C. R., and Agostinho, P. (2008). Amyloid-beta peptide decreases glutamate uptake in cultured astrocytes: Involvement of oxidative stress and mitogen-activated protein kinase cascades. *Neuroscience* **156,** 898–910.

Meyer-Luehmann, M., Spires-Jones, T. L., Prada, C., Garcia-Allosa, M., de Calignon, A., Rozkalne, A., Koeningsknecht-Talboo, J., Holtzman, D. M., Bacskai, B. J., and Hyman, B. T. (2008). Rapid appearance and local toxicity of amyloid-beta plaques in a mouse model of Alzheimer's disease. *Nature* **451,** 720–724.

Nizzari, M., Venezia, V., Repetto, E., Caorsi, V., Magrassi, R., Gagliani, M. C., Carlo, P., Florio, T., Schettini, G., Tacchetti, C., Russo, T., Diaspro, A., *et al.* (2007). Amyloid precursor protein and presenilin 1 interact with the adaptor GRB2 and modulates ERK 1,2 signaling. *J. Biol. Chem.* **282,** 13833–13844.

Oddo, S., Caccamo, A., Green, K. N., Liang, K., Tran, L., Chen, Y., Leslie, F. M., and LaFerla, F. M. (2005). Chronic nicotine administration exacerbates tau pathology in a transgenic model of Alzheimer's disease. *Proc. Natl. Acad. Sci. USA* **102,** 3046–3051.

Papassotiropoulos, A., Lambert, J. C., Wavrant-De Vrieze, F., Wollmer, M. A., von der Kammer, H., Streffer, J. R., Maddalena, A., Huynh, K. D., Wolleb, S., Lutjohann, D., Schneider, B., Thal, D. R., *et al.* (2005). Cholesterol 25-hydroxylase on chromosome 10q is a susceptibility gene for sporadic Alzhmeimer's disease. *Neurodegener. Dis.* **2,** 233–241.

Permanne, B., Adessi, C., Saborio, G. P., Fraga, S., Frossard, M. J., Van Dorpe, J., Dewachter, I., Banks, W. A., Van Leuven, F., and Soto, C. (2002). Reduction of amyloid load and cerebral damage in transgenic mouse model of Alzheimer's disease by treatment with a beta-sheet breaker peptide. *FASEB J.* **16,** 860–862.

Pigino, G., Morfini, G., Pelsman, A., Mattson, M. P., Brady, S. T., and Busciglio, J. (2003). Alzheimer's presenilin 1 mutations impair kinesin-base axonal transport. *J. Neurosci.* **23,** 4499–4508.

Rajendra, L., Schneider, A., Schlechtingen, G., Weidlich, S., Ries, J., Braxmeier, T., Schwille, P., Schulz, J., Schroeder, C., Simons, M., Jennings, G., Knolker, H. J., *et al.* (2008). Efficient inhibition of the Alzheimer's disease beta-secretase by membrane targeting. *Science* **320,** 520–523.

Roberson, E. D., Scearce-Levie, K., Palop, J. J., Yan, F., Cheng, I. H., Wu, T., Gerstein, H., Yu, G. Q., and Mucker, L. (2007). Reducing endogenous tau ameliorates amyloid b-induced deficits in an Alzheimer's disease mouse model. *Science* **316,** 750–754.

Rogaev, E. I., Sherrington, R., Rogaeva, E. A., Levesque, G., Ikeda, M., Liang, Y., Chin, H., Lin, C., Holman, K., Tsuda, T., Mar, L., Sorbi, S., *et al.* (1995). Familial Alzheimer's disease in kindreds with missense mutations in a gene on chromosome 1 related to Alzheimer's disease type 3 gene. *Nature* **376,** 775–778.

Rossor, M. N., Svendsen, C., Hunt, S. P., Mountjov, C. Q., Roth, M., and Iversen, L. L. (1982). The substantia innominata in Alzheimer's disease: An histochemical and biochemical study of cholinergic marker enzymes. *Neurosci. Lett.* **28,** 217–222.

Sadowski, M., Pankiewicz, J., Scholtzova, H., Ripellino, J. A., Li, Y., Schmidt, S. D., Mathews, P. M., Fryer, J. D., Holtzman, D. M., Sigurdsson, E. M., and Wisniewski, T. (2004). A synthetic peptide blocking the apolipoprotein E/beta-amyloid binding mitigates beta-amyloid toxicity and fibril formation *in vitro* and reduces beta-amyloid plaques in transgenic mice. *Am. J. Pathol.* **165,** 937–948.

Saito, K., Elce, J. S., Hamos, J. E., and Nixon, R. A. (1993). Widespread activation of calcium-activated neutral proteinase (calpain) in the brain in Alzheimer's disease: A potential molecular basis for neuronal degeneration. *Proc. Natl. Acad. Sci. USA* **90,** 2628–2632.

Sarmay, G., Angyal, A., Kertesz, A., Maus, M., and Medgyesi, D. (2006). The multiple function of Grb2 associated binder (Gab) adaptor/scaffolding protein in immune cell signaling. *Immunol. Lett.* **104,** 76–82.

Scapagnini, G., Clombrita, C., Amadio, M., D'Agata, V., Arcelli, E., Sapienza, M., Quattrone, A., and Calabrese, V. (2006). Curcumin activates defensive genes and protects neurons against oxidative stress. *Antioxid. Redox. Signal.* **8,** 395–403.

Schilling, S., Zeitschel, U., Hoffmann, T., Heiser, U., Francke, M., Kehlen, A., Holzer, M., Hutter-Paier, G., Prokesch, M., Windisch, M., Jagla, W., Schlenzig, D., *et al.* (2008). Glutaminyl cyclase inhibition attenuates pyroglutamate Aβ and Alzheimer's disease-like pathology. *Nat. Med.* **14,** 1106–1111.

Schwarze-Eicker, K., Keyvani, K., Gortz, N., Westaway, D., Sachser, N., and Paulus, W. (2005). Prion protein (PrPc) promotes beta-amyloid plaque formation. *Neurobiol. Aging* **26,** 1177–1182.

Selkoe, D. J. (2001). Alzheimer's disease: Genes, proteins, and therapy. *Physiol. Rev.* **81,** 741–766.

Sherrington, R., Rogaev, E. I., Liang, Y., Rogaeva, E. A., Levesque, G., Ikeda, M., Chi, H., Lin, C., Li, G., Holman, K., Tsuda, T., Mar, L., *et al.* (1995). Cloning of a gene bearing missence mutations in early-onset familial Alzheimer's disease. *Nature* **375,** 754–760.

Sisodia, S. S., and St George-Hyslop, P. H. (2002). Gamma-secretase, notch, Abeta and Alzheimer's disease: Where do the presenilins fit in? *Nat. Rev. Neurosci.* **3,** 281–290.

Suh, Y. H., and Checler, F. (2002). Amyloid precursor protein, presenilins, and alpha-synuclein: Molecular pathogenesis and pharmacologal applications in Alzheimer's Disease. *Pharmacol. Rev.* **54,** 469–525.

Tai, H. C., and Schuman, E. M. (2008). Ubiquitin, the proteasome and protein degradation in neuronal function and dysfunction. *Nat. Rev. Neurosci.* **9,** 826–838.

Wolfe, M. S., Xia, W., Ostaszewski, B. L., Diehl, T. S., Kimberly, W. T., and Selkoe, D. J. (1999). Two transmembrane aspartates in presenilin-1 required for presenilin endoproteolysis and gamma-secretase activity. *Nature* **398,** 513–517.

Wu, J., Kuo, Y. P., George, A. A., Xu, L., Hu, J., and Lukas, R. J. (2004). Beta- amyloid directly inhibits human alpha 4 beta 2 nicotinic acetylcholine receptors heterologously expressed in human SH-EP1 cells. *J. Biol. Chem.* **279,** 37842–37851.

Yamamoto, A., Shin, R. W., Hasegawa, K., Naiki, H., Sato, H., Yoshimasu, F., and Kitamoto, T. (2002). Iron (III) induces aggregation of hyperphosphorylated tau and its reduction to iron (II) reverses the aggregation: Implications in the formation of neurofibrillary tangles of Alzheimer's disease. *J. Neurochem.* **82,** 1137–1147.

Yamazaki, H., Bujo, H., Kusonoki, J., Seimiya, K., Kanaki, T., Morisaki, N., Scheneider, W. J., and Saito, Y. (1996). Elements of neural adhesion molecules and a yeast vacuolar protein sorting receptor are present in a novel mammalian low density lipoprotein receptor family member. *J. Biol. Chem.* **271,** 24761–24768.

Younkin, S. G. (1998). The role of A beta 42 in Alzheimer's disease. *J. Physiol. (Paris)* **92,** 289–292.

Yu, G., Niashimura, M., Arawaka, S., Levitan, D., Zhang, L., Tandon, A., Song, Y. Q., Rogaeva, E., Chen, F., Kawarai, T., Supala, A., Levesque, L., *et al.* (2000). Nicastrin modulates presenilin mediated notch/glp-1 signal transduction and betaAPP processing. *Nature* **407,** 48–54.

Zhang, Q., Powers, E. T., Nieva, J., Hyff, M. E., Dendle, M. A., Bieschle, J., Glabe, C. G., Eschenmoser, A., Wentworth, P., Lerner, R. A., and Kelly, J. W. (2004). Metabolite-initiated protein misfolding may trigger Alzheimer's disease. *Proc. Natl. Acad. Sci. USA* **101,** 4752–4757.

ALZHEIMER'S DISEASE GENETICS: CURRENT STATUS AND FUTURE PERSPECTIVES

Lars Bertram

Neuropsychiatric Genetics Group, Department of Vertebrate Genomics, Max-Planck Institute for Molecular Genetics, Berlin 14195, Germany
Genetics and Aging Research Unit, Department of Neurology, Massachusetts General Hospital, Charlestown, Massachusetts, 02129, USA

Alzheimer's disease (AD) is a genetically complex disease whose pathogenesis is largely influenced by genetic factors. Three decades of intensive research have yielded four established AD genes (*APP*, *PSEN1*, *PSEN2*, *APOE*), and hundreds of potential susceptibility loci, none of which has been unequivocally shown to modify disease risk using conventional methodologies. The results of genome-wide association studies (GWAS) are now adding to an already vast and complicated body of data. To facilitate the evaluation and interpretation of these findings, we have recently created a database for genetic association studies in AD ("AlzGene"; available at http://www.alzgene.org). In addition to systematically screening and summarizing the scientific literature for eligible studies, AlzGene provides the results of allele-based meta-analyses for all polymorphisms with sufficient genotype data. Currently, these meta-analyses highlight over 20 different potential AD genes, several of which were originally implicated by a GWAS. First follow-up analyses in a large collection of over 1300 AD families reveal that—in addition to *APOE*—genetic variants in *ACE*, *CHRNB2*, *GAB2*, and *TF* show the most consistent risk effects across a wide range of independent samples and study designs. The chapter highlights these and other promising findings from the recent AD genetics literature and provides an overview of the powerful new tools aiding researchers today to unravel the genetic underpinnings of this devastating disease.

INTERNATIONAL REVIEW OF NEUROBIOLOGY, VOL. 84
DOI: 10.1016/S0074-7742(09)00409-7

I. Introduction: The Genetic Dichotomy of Alzheimer's Disease

Similar to many other adult-onset disorders, Alzheimer's disease (AD) shows a genetic dichotomy of cases with Mendelian inheritance ("simplex AD") versus cases governed by an array of different genetic and nongenetic risk factors ("complex AD"; Bertram and Tanzi, 2005; Tanzi, 1999). Simplex AD is rare (<5%) and usually shows an early (<65 years) or often times very early (<50 years) age of onset and, owing to its typical familial accumulation and inheritance pattern, is also known as "early-onset familial AD" (EOFAD). Currently, mutations in three genes are known to cause EOFAD, all leading to altered cerebral levels of the amyloid-β peptide (Aβ), the main constituent of senile plaques, and one of the neuropathological hallmarks of the disease (Tanzi and Bertram, 2005). The mutated genes leading to EOFAD are APP (amyloid precursor protein; Goate *et al.*, 1991), PSEN1 (presenilin 1; Sherrington *et al.*, 1995), and PSEN2 (presenilin 2; Levy-Lahad *et al.*, 1995; Rogaev *et al.*, 1995), and all three are intimately involved in the production of Aβ, which is liberated from APP via two enzymatic cleavage events, mediated by β- and γ-secretase. The latter is a protein complex consisting of various components, in which the presenilins form the γ-secretase active site. Even though EOFAD represents only a tiny fraction of all AD cases, the knowledge gained from uncovering its genetic underpinnings has been instrumental in our current understanding of the forces leading to AD across all onset ages and have corroborated the "amyloid hypothesis." This hypothesis posits that a dysregulation of the production/function of Aβ is the initiating event in the pathogenetic cascade eventually leading to neurodegeneration and dementia (Glenner and Wong, 1984; reviewed in Hardy and Selkoe, 2002; Tanzi and Bertram, 2005). Finally, many of the currently most promising therapeutic approaches in the AD field aim to directly intersect or reverse the pathological build-up of Aβ, which for the first time would allow a causal treatment of AD as opposed to the merely symptomatic options available to date.

Despite its relatively high heritability (Bergem *et al.*, 1997; Gatz *et al.*, 2006; Meyer and Breitner, 1998)—that is, the proportion by which phenotypic variation is determined by genetic variation—the genetic make-up of complex AD, which, owing to its later onset age ($\geq$ 65 years), is also known as "late-onset AD" (LOAD), has proved to be much more difficult to decipher. This is at least in part due to problems that generally aggravate epidemiological research in many neuropsychiatric diseases, for example, small genetic effect sizes and reduced penetrance of the involved genetic factors, a considerable degree of phenotypic variability and co-morbidity with other late-onset disorders, the lack of extended pedigrees for LOAD cases, and the absence of disease-specific biomarkers (Kennedy *et al.*, 2003). The identification of susceptibility genes is further complicated by gene-gene interactions that are difficult to predict and model, and a sizeable but difficult to detect and measure environmental component. Notwithstanding these challenges, several chromosomal regions thought to harbor novel LOAD genes have been identified via whole genome linkage analyses, some, for example, on chromosomes 9, 10, and 12, overlapping across different samples (Bertram *et al.*, 2000; Blacker *et al.*, 2003; Ertekin-Taner *et al.*, 2000; Hamshere *et al.*, 2007; Kehoe *et al.*, 1999; Li *et al.*, 2002; Myers *et al.*, 2000; Pericak-Vance *et al.*, 2000). Furthermore, well over 1000 "candidate gene" studies—that is, studies that focus on certain genes based on some prior hypothesis regarding their potential involvement in the disease process—have been published over the past two decades claiming or refuting genetic association between putative AD genes and affection status and/or certain endophenotypes (Bertram *et al.*, 2007). Currently, nearly 150 AD genetic association papers are published each year, at increasing pace. However, with the exception of one genetic variant, the ε4-allele of the apolipoprotein E gene (*APOE*; Saunders *et al.*, 1993; Strittmatter *et al.*, 1993) none of these candidates has been proved to consistently influence disease risk or onset age in more than a handful of samples (Bertram and Tanzi, 2008). Instead, most studies of "novel AD genes" have been followed by a large number of conflicting reports, challenging prior claims that they may play an important role in contributing to disease risk. For geneticists as well as clinicians the growing number of (mostly conflicting) genetic association findings in AD has become increasingly difficult to follow, evaluate, and interpret.

II. Genome-Wide Association Studies in AD

An alternative to the traditional candidate gene approach is afforded by recent advances in large-scale genotyping technologies which now enable researchers to perform comprehensive and largely hypothesis-free genome-wide association studies (GWAS; Craddock *et al.*, 2008; McCarthy *et al.*, 2008). Four groups have reported the results of AD GWAS to date (Table I; Bertram *et al.*, 2008a; Coon *et al.*, 2007; Grupe *et al.*, 2007; Li *et al.*, 2008). The first (Grupe *et al.*, 2007) tested roughly 17,000 single nucleotide polymorphisms (SNPs) in or very near genetic coding regions ("cSNPs"). The only polymorphisms found to consistently associate with AD risk across the different samples were located in close proximity to *APOE*, and most likely reflect linkage disequilibrium (LD) with the ε4-allele. Although a number of additional loci were highlighted as potential AD genes by the authors, none showed the same consistency of effect or same level of statistical significance as the ε4-related variants. The second study (Coon *et al.*, 2007) tested ~500,000 SNPs in roughly 1100 unrelated AD cases and controls.

TABLE I

OVERVIEW OF ALL PUBLISHED GWAS IN AD

GWA study	Design	Population	No. SNPs	No. AD cases (total)[a]	No. controls (total)[a]	"Featured" genes
Grupe *et al.* (2007)	Case-control	USA and UK	17,343	380 (1428)	396 (2062)	*APOE*, ACAN, BCR, CTSS, EBF3, GALP, GWA_14q32.13, GWA_7p15.2, LMNA, LOC651924, MYH13, PCK1, PGBD1, THEM5, TNK1, TRAK2, UBD*
Coon *et al.* (2007) and Reiman *et al.* (2007)	Case-control	USA, Netherlands	502,627	446 (861)	290 (550)	*APOE*, GAB2*
Li *et al.* (2008)	Case-control	Canada	469,438	753 (1071)	736 (985)	*APOE*, GOLM1, GWA_15q21.2, GWA_9p24.3*
Bertram (2008)	Family-based	USA	484,522	941 (2708)	404 (1242)	*APOE*, ATXN1, CD33, GWA_14q31,*

Modified after content on the AlzGene Web site (http://www.alzgene.org; current on October 31st, 2008). Studies are listed in order of publication date. "Featured Genes" are those genes/loci that were declared as "associated" in the original publication, note that criteria for declaring association may vary across studies.

* Indicates that surrogate markers were used for *APOE*.

[a]Numbers of "AD cases" and "controls" refer to sample sizes used in initial GWA screening, whereas "total" refers to initial sample plus any follow-up samples (where applicable); please consult AlzGene Web site for more details on these studies.

Again, with the exception of a single SNP in strong LD with *APOE*-ε4, no other genome-wide significant signals were observed. In a follow-up paper (Reiman *et al.*, 2007), the same group reported evidence of association with variants in the gene encoding GRB2-associated binding protein 2 (*GAB2*) on chromosome 11q14, which only became evident after their samples were stratified for *APOE*-ε4, that is, when AD cases and controls were divided into carriers and non-carriers of the ε4-allele. A third group (Li *et al.*, 2008) tested nearly ~470,000 markers and reported association with variants in golgi membrane protein 1 (*GOLM1*) on chromosome 9q22 and two currently uncharacterized loci on chromosomes 9p and 15q. Finally, our own group (Bertram *et al.*, 2008a) recently reported the results of the first GWAS using family-based samples (~1400 DNAs from over 400 families), as opposed the three previously summarized studies which were solely based on case-control datasets. Applying the same 500 K marker array as in Coon *et al.* (2007) and Li *et al.* (2008) we found—in addition to *APOE*-ε4—evidence for a number of potential AD loci (Table I). Although the findings from these first GWAS hold the promise of pinpointing novel pathways and mechanisms potentially important in AD pathogenesis, it will be important to await consistent replication by independent investigators.

While these and several of the forthcoming GWAS have the potential to significantly advance our understanding of the genetics and pathogenetic mechanisms of AD and other neurodegenerative disorders, it needs to be emphasized that in many ways genome-wide screens are actually not so different from conventional candidate gene association analyses. Both search for significantly different allele or genotype distributions or transmissions in subjects affected by the disease/phenotype as compared to presumably healthy individuals. The two approaches differ mostly on a *quantitative* level: instead of testing a few tens of markers (or less), GWAS simultaneously screen a few hundreds of thousands of markers (or more). The major *qualitative* difference between a GWAS and candidate gene analysis is that the former investigates the whole genome in a more or less unbiased fashion, whereas the latter only investigates a limited number of specific loci proved or thought to be involved in disease predisposition or progression (e.g., in AD many of these loci are involved in the production, trafficking, or removal of Aβ). On the downside and owing to their sheer number of polymorphisms tested, GWAS actually substantially compound the problem that has plagued genetic studies of complex phenotypes in the past, that is, to determine which of the many reported putative risk alleles are "real" as opposed to which merely reflect statistical artifacts. The first essential step in differentiating these two alternatives is to provide independent replication of the association (McCarthy *et al.*, 2008), just as for any result emerging from "old-fashioned" candidate gene analyses.

III. Systematic Field Synopsis and Meta-Analyses: The AlzGene Database

In an attempt to facilitate the interpretation of association findings regardless of the technology used for initial detection, we have recently created a publicly available database, "AlzGene" (http://www.alzgene.org), which systematically collects, summarizes and meta-analyzes all genetic association studies published in the field of AD, including GWAS (Bertram *et al.*, 2007). After thorough (and ongoing) searches of the available scientific literature, studies published in peer-reviewed journals that are available in English are included in the database. Key variables (such as ancestry, type of AD diagnosis, sample size, onset age, and genotype distributions) are extracted from the original publications. Furthermore, published genotype data from independent case-control samples are systematically meta-analyzed. Because this approach quantitatively synthesizes all of the published genotype data for each polymorphism, it facilitates the overall interpretation of association findings: rather than relying on the either "positive" (that is, statistically significant) or "negative" (that is, insignificant) outcomes of individual studies, it produces a summary risk estimate (called the odds ratio) that takes into consideration all data at once and accounts for within and between study variation.

Currently, AlzGene includes more than 1100 individual studies and showcases the results of over 200 individual meta-analyses. In these, more than 20 genes that are not related to the well-established *APOE*-ε4 allele show nominally significant risk effects (Table II). Interestingly, about one fourth of these were originally implicated by one of the four published GWAS, although independent replication by other groups is still lacking for most of these findings. As can be seen in Table II, the average allelic summary odds ratio (OR), for non-*APOE*-related effects are very modest (~1.2 for "risk" alleles and ~0.8 for "protective" alleles) compared to an OR of ~3–4 for a single copy of the *APOE*-ε4 allele. These modest effect sizes are in good agreement with those found in other large-scale studies on the genetics of complex diseases (Allen *et al.*, 2008; Ioannidis *et al.*, 2001; Lohmueller *et al.*, 2003) and have important (and well known) implications for the design of future genetic association studies in AD, as sample sizes will need to be vastly increased to detect or exclude ORs of 1.25 or below with sufficient confidence. For instance, to detect an allelic OR of 1.25 with 80% power at a *P*-value of 0.05 sample sizes between ~1400 and 6000 combined cases and controls are needed for disease allele frequencies ranging from 0.50 to 0.05, respectively (Bertram, 2008). Sample sizes need to be increased approximately fivefold to detect such modest effects with the same power at *P*-values below 5×10^{-8}, that is, one proposed threshold for declaring genome-wide significance (Hoggart *et al.*, 2008; McCarthy *et al.*, 2008).

TABLE II

ALZGENE "TOP RESULTS" (CURRENT ON OCTOBER 31ST, 2008)

Locus/Gene	Polymorphism	AlzGene OR (95% CI)[a]	*P*-value	No. AD	No. controls	No. Samples[b]	Ethnicity
APOE[c]	ε2/3/4	3.81 (3.38-4.29)	$<1 \times 10^{-30}$	2325	3574	23	CAU
CHRNB2	rs4845378	0.67 (0.50-0.90)	0.007	576	787	4	ALL
GAB2	rs10793294	0.69 (0.54-0.88)	0.0025	2073	1836	5	CAU
CH25H	rs13500	1.44 (1.08-1.93)	0.01	1549	1506	6	ALL
SORL1	SORL1-18ex26	0.70 (0.50-0.98)	0.04	743	913	3	CAU
CALHM1	rs2986017	1.42 (1.27-1.59)	1.2×10^{-9}	2043	1361	5	ALL
CST3	5′ UTR-157	1.28 (1.05-1.56)	0.02	804	571	3	CAU
ACE	rs1800764	0.79 (0.68-0.92)	0.002	818	747	4	CAU
PGBD1	rs3800324	1.25 (1.04-1.51)	0.02	2736	2964	6	ALL
MAPT/STH	rs2471738	1.24 (1.01-1.53)	0.04	1734	1411	6	ALL
SORCS1	rs600879	1.24 (1.04-1.48)	0.02	1361	1495	4	ALL
hCG2039140	rs1903908	1.23 (1.06-1.44)	0.007	1368	1497	4	ALL
TFAM	rs2306604	0.82 (0.72-0.94)	0.003	1059	792	5	ALL
IL1B	rs1143634	1.18 (1.03-1.34)	0.02	1011	1244	5	ALL
TF	rs1049296	1.18 (1.04-1.33)	0.01	1916	4058	9	ALL
TNK1	rs1554948	0.86 (0.80-0.93)	0.0002	2743	2984	6	ALL
LOC439999	rs498055	1.15 (1.03-1.29)	0.02	2787	2498	7	ALL
GAPDHS	rs4806173	0.87 (0.75-1.00)	0.05	1687	1775	4	ALL
DAPK1	rs4878104	0.88 (0.82-0.95)	0.002	2753	3036	7	ALL
PRNP	rs1799990	0.88 (0.81-0.96)	0.03	2280	2943	10	CAU
GWA_14q32.13	rs11622883	0.88 (0.80-0.97)	0.01	2685	2893	6	ALL
GALP	rs3745833	1.13 (1.00-1.29)	0.05	2739	2975	6	ALL

(*Continued*)

TABLE II (*Continued*)

Locus/Gene	Polymorphism	AlzGene OR (95% CI)[a]	*P*-value	No. AD	No. controls	No. Samples[b]	Ethnicity
LOC651924	rs6907175	0.89 (0.82-0.96)	0.005	2359	2522	6	ALL
NEDD9	rs760678	0.89 (0.81-0.97)	0.01	3452	3245	8	ALL
MTHFR	rs1801133	1.11 (1.02-1.21)	0.02	2887	4663	24	ALL
ENTPD7	rs911541	1.10 (1.01-1.21)	0.03	3429	3743	4	ALL
BDNF	rs6265	1.09 (1.02-1.18)	0.015	5299	4671	15	CAU
IL1A	rs1800587	1.09 (1.00-1.18)	0.04	5296	5321	25	ALL

List of loci, in descending order of genetic effect size, containing at least one polymorphism showing nominally significant (*P*-value ≤ 0.05) summary ORs in AlzGene on October 31st, 2008. To be considered as "Top Result," summary OR needs to be significant across samples from all ethnic backgrounds ("ALL") or in Caucasians only ("CAU"). Whenever nominally statistically-significant results are observed for both, that is, ALL and CAU, only the analysis that has the largest genetic effect size (OR deviating the most from 1) is reported here. Note that AlzGene is continuously updated, so results displayed online may differ from the results above; consult the AlzGene Web site for up-to-date numbers and additional meta-analyses in these and other loci (http://www.alzgene.org) (see also Table I).

[a]Summary ORs and 95% confidence intervals (CI) are based on random-effects allelic contrasts comparing minor versus major alleles at each polymorphism.,

[b]Number of samples refers to the number of independent case-control samples used in the meta-analyses; multiple samples may be reported in the same publication and are considered separately if they are independent, that is, non-overlapping. Samples overlapping across publications are only used once, usually by including the datasets with the largest number of available genotypes,

[c]Results are based on comparing $\varepsilon 4$- versus $\varepsilon 3$-alleles at this locus.

IV. Strengths and Limitations of the AlzGene Approach

The strengths of AlzGene are obvious: assuming that the literature searches, inclusion criteria, data management, and data analysis procedures actually provide a correct and exhaustive account of the available literature, AlzGene is the single most comprehensive resource for the status of genetics research in AD available to date. In our original data freeze (Bertram *et al.*, 2007) we could show that literature searches in AlzGene outperformed those of several other literature/genetics databases, and that the results of our meta-analyses were in very good agreement with estimates published previously in nearly two dozen individual papers. Similar observations were made in related database projects from our group (Allen *et al.*, 2008; Bagade *et al.*, 2006). Published meta-analyses, however, have one important disadvantage: by nature of their design, they run the risk of becoming outdated quickly, possibly as soon as new data from one or two additional studies are published. Provided that sufficient funding remains available, AlzGene does not have this caveat. Also, any meta-analysis in the database can be updated literally within days after the publication of novel data. Another strength of AlzGene is that it is not limited to meta-analyses on certain genes or networks of genes (e.g., those that are in the same pathway or gene family), but considers all published loci simultaneously, making the comparison of results across studies, genes, pathways, chromosomal regions, and so on extremely easy. Finally, as outlined above, all loci containing at least one polymorphism nominally significant by meta-analysis are separately highlighted on the database's homepage in a section called "Top Results." Thus, consulting this section of the AlzGene Web site will provide the user with a complete—and essentially real time—snapshot of the "most promising" AD candidate genes, based on the systematic evaluation of literally hundreds of individual studies and thousands of data points. As such, the "Top Results" list could help prioritize future genetic association studies (e.g., for further independent replication, or fine mapping), and guide functional genetics and molecular studies investigating the potential pathogenetic mechanisms underlying the genetic associations.

While AlzGene undoubtedly represents a leap forward in managing and displaying the data gathered within the domain of AD genetics research, its overall approach, too, comes with some strings attached. First and foremost, despite our comprehensive and systematic searches of the scientific literature, we cannot exclude the possibility that some AD association studies were overlooked or entered erroneously. This can be partly alleviated with the help of database users who are explicitly encouraged to alert the curatorial team of any errors or omissions, which will be fixed as soon as possible. Other limitations include our restriction to allele contrasts in the meta-analyses (which allows no inference of the true underlying mode of inheritance and is usually less powerful than genotype-based tests), the non-consideration of haplotype-based genotype data or imputed single-locus genotypes (possibly missing important associations), the exclusive focus on "main effects" (and the inherent inability to account for gene-gene and gene-environment interactions), and the lack of adjustment for certain co-variates such as age or gender (which is impossible to do systematically without access to subject-level raw data). Further, protection from bias, in particular publication bias and other sorts of reporting biases, is particularly difficult to ensure or assess, and is likely to affect some of the meta-analysis results. Finally, the single most important caveat of the AlzGene approach is that the number of "true" associations is almost certainly going to be smaller than the number of nominally significant findings listed at any time on the AlzGene Web site (see also: Ioannidis, 2005; Wacholder *et al.*, 2004). This has a number of causes, including multiple testing, linkage disequilibrium among associated variants, undetected publication or other reporting biases, as well as study-level technical artifacts that may have gone unnoticed or may be impossible to detect. Furthermore, most of the "positive" meta-analysis outcomes currently do not reach very high levels of statistical significance (see Table II), only two attaining the threshold for genome-wide significance, for example, a *P*-value $\leq 5 \times 10^{-8}$. While this, too, could be due to a number of factors including small effect size and insufficient power even after combining all of the available data, it is important to emphasize that the possibility of a false-positive finding always exists, even for the highest ranked "Top Result." Eventually, genuine risk effects can only be "proved" by the accumulation of sufficient unbiased and high-quality genotype data (e.g., originating from case-control *and* family-based designs, see below) in favor of the presumed association in combination with functional genetics and biological evidence suggesting a direct biochemical effect of the associated variant (McCarthy *et al.*, 2008). Of course, such evidence can be difficult to come by. For instance, despite the unambiguous role of the *APOE*-ε4 allele in increasing the risk for AD (see above), the precise mechanism underlying this association remains only poorly understood (Tanzi and Bertram, 2005). Notwithstanding these limitations, there is good reason to believe that the variants and loci highlighted in the "Top Results" section of AlzGene and related databases currently represent our best bets as to which of the hundreds of putative candidate genes might genuinely contribute to disease susceptibility and pathogenesis. As such, they probably warrant follow-up with high priority.

V. Assessment of Case-Control Associations Using Family-Based Methods

A first round of genetic follow-up of the most promising of these associations was recently completed in a number of projects from our group (Bertram *et al.*, 2008b; Schjeide *et al.*, 2009a,b). In particular, we assessed whether or not any of the AlzGene "Top Results" and/or any of the loci pinpointed by the two high-density case-control GWAS (Li *et al.*, 2008; Reiman *et al.*, 2007) showed association in a total of four family-based datasets comprising nearly 4200 DNAs from over 1300 independent AD families. This was an important question as many of the investigated variants had been thoroughly tested across relatively large numbers of independent case-control datasets, but only a handful were also previously assessed in family-based samples, which may be genetically different from unrelated, population-based cases and controls. Family-based methods have the additional advantage of being robust against bias due to undetected population stratification and phenotype misspecifications (Laird and Lange, 2006), which may have affected some of the case-control meta-analysis results. After combining the results across all four datasets, we observed nominally significant association with variants in *ACE*, *CHRNB2*, *GAB2*, *TF*, and an as yet unidentified locus on chromosome 7p15.2 (Table III). Two of these loci, that is, *ACE* and *TF*, were also found to be associated with Aβ levels in CSF in an independent collection of AD cases and controls (Kauwe *et al.*, 2009). The independent convergence of (i) significant meta-analysis results across case-control samples, (ii) replication of these associations in AD family samples, and (iii) in the case of *ACE* and *TF*, a significant genotype-dependent correlation with one of the few established biomarkers in AD, strongly implies a genuine disease-risk modifying role of these loci, arguably more so than for any of the other hundreds of suggested AD candidate genes besides *APOE*.

Functionally, all of these potential new AD susceptibility loci are interesting and, if confirmed, may not only lead to a better understanding of the pathogenetic mechanisms driving neurodegeneration, but may also help to pinpoint novel treatment options for AD. *ACE* encodes angiotensin converting enzyme-1 (ACE-1), a ubiquitously expressed zinc metalloprotease that is involved in blood pressure regulation. Several epidemiological studies suggest that high mid-life blood pressure may increase the risk for AD in later life (Takeda *et al.*, 2008). More recently, ACE-1 activity has been reported to be increased in AD brains proportionately to parenchymal Aβ load (Miners *et al.*, 2008). The interpretation of ACE's role in AD pathogenesis is complicated by the observation that it is able to degrade naturally secreted Aβ *in vitro* (Hu *et al.*, 2001), which would predict an increased risk in individuals with reduced ACE-1 levels/activity, that is, opposite to what would be expected from the epidemiological data and the genetic association results. *CHRNB2* encodes the neuronal beta-2 nicotinic cholinergic receptor (β2nAChR). Nicotinic acetylcholine receptors (nACcRs) are widely expressed in the CNS, where the β2nAChR subunit is particularly abundant, forming hetero-pentameric receptors with β4nAChR subunits ($\alpha 4\beta 2$; Kalamida *et al.*, 2007). Pathologically, the reduction of nAChRs and the loss of cholinergic neurons in disease-relevant brain regions is one of the major neurochemical hallmarks of AD (Oddo and LaFerla, 2006), and several studies have suggested that an age-dependent decrease in protein and/or mRNA levels of the $\alpha 4\beta 2$-subtype (in particular β2nAChR) occurs in the cortex and hippocampus of healthy

TABLE III

ASSESSMENT OF CASE-CONTROL ASSOCIATIONS IN AD FAMILY DATASETS

				Case-control (AlzGene)		Family based (combined family datasets)			
Gene	SNP	Ethnicity	Model	OR (95% CI)	P-value	No. Fams	OR	χ^2	*P*-value[a]
ACE	rs4291	CAU	T vs. A	0.87 (0.76-0.99)	0.03	425	0.94	14.6	0.07
CHRNB2	rs4845378	ALL	T vs. G	0.67 (0.50-0.90)	0.007	170	0.79	17.4	0.03
GAB2	rs7101429	CAU	G vs. A	0.74 (0.59-0.93)	0.008	432	0.76	24.4	0.002
GWA_7p15.2	rs1859849	ALL	C vs. T	1.11 (0.97-1.29)	0.1	335	1.26	17.5	0.03
TF	rs104296	ALL	C2 vs. C1	1.18 (1.04-1.33)	0.01	295	1.17	21.1	0.007

Comparison of the association evidence of AlzGene "Top Results" (current on October 31, 2008) and family-based analyses (Schjeide *et al.*, 2009a,b). Note that the meta-analyses for rs1859849 in GWA_7p15.2 are currently only marginally significant (*P*-value 0.1), but were nominally significant at the time of genotyping in the Schjeide *et al.* (2009a) project. For up-to-date meta-analysis results on these variants, please consult the AlzGene Web site (http://www.alzgene.org). ALL = combining samples of all ethnic groups; CAU = combining samples of Caucasian ethnicity only; No. Fams = No. informative families; "OR" = summary odds ratio; "95% CI" = 95% confidence intervals.

[a]Test statistics are based on combining one-tailed *P*-values for each variant across all four samples using Fisher's combined probability test, which results in an 8 d.f. test. Single-locus analyses with variants in *ACE* only show marginally significant effects (*P*-value 0.07), while haplotype-based analyses yielded *P*-values <0.05 (see Schjeide *et al.*(2009a) for more details).

individuals (Tohgi *et al.*, 1998), which could be exacerbated in carriers of the AD risk allele. *GAB2* encodes GRB2-associated binding protein 2 (Gab2) which is a member of a family of evolutionarily highly conserved scaffolding/adapter proteins that are involved in multiple signaling pathways, and in particular in the transduction of cytokine and growth receptor signaling (Liu and Rohrschneider, 2002; Sarmay *et al.*, 2006). Gab2 is ubiquitously expressed, but is found at particularly high levels in white blood cells, prefrontal cortex, and hypothalamus. It was suggested that changes in Gab2 expression could potentially affect Gsk3-dependent phosphorylation of tau and the formation of neurofibrillary tangles (Reiman *et al.*, 2007). Moreover, growth factor receptor-bound protein 2, which binds Gab2, has been reported to bind both tau (Reynolds *et al.*, 2008), APP, and presenilin 1 and 2 (Nizzari *et al.*, 2007), and these interactions have been proposed to regulate signal transduction. *TF* encodes transferrin which is the major circulating glycoprotein involved in iron metabolism and is highly expressed in the brain. There is a vast body of literature suggesting that iron dysregulation promotes neurodegeneration, possibly via the generation of reactive oxygen species (Brewer, 2007). In AD, iron has been found to be increased in the brains of AD patients (Loeffler *et al.*, 1995), where it is associated with plaques and NFTs (Smith *et al.*, 1997). More recently, it was suggested that iron may also play a role in the aggregation of hyperphosphorylated tau into insoluble paired helical filaments, one of the core ingredients of NFTs (Yamamoto *et al.*, 2002). Thus it is conceivable, although not yet experimentally proved, that the AD-associated amino acid changing variant in *TF* (Pro570Ser) may interfere with all or some of these roles in AD pathogenesis. Finally, the *chromosome 7p15.2* association signal maps close to a predicted gene, NT_007819.514, encoding a protein of 358 residues, whose N-terminus exhibits a strong homology to a family of ubiquitin-like proteins, for example, human ubiquitin C. Thus, the predicted protein possibly plays a role in protein degradation.

Figure 1 summarizes the potential pathogenetic implications of these and several other of the currently most compelling AD susceptibility genes. Note that this summary represents only a snapshot of the presently available data, and that the set of genes thought to be involved in predisposing for LOAD today is likely going to change over time. For a more detailed review on the functional implications of these and other compelling AD candidate genes see Bertram and Tanzi (2008).

VI. Conclusions

Despite intensifying efforts to unravel the genetic underpinnings of AD, successes to date have been modest. This situation is expected to change dramatically with the advent of novel, genome-wide analysis tools that are becoming increasingly

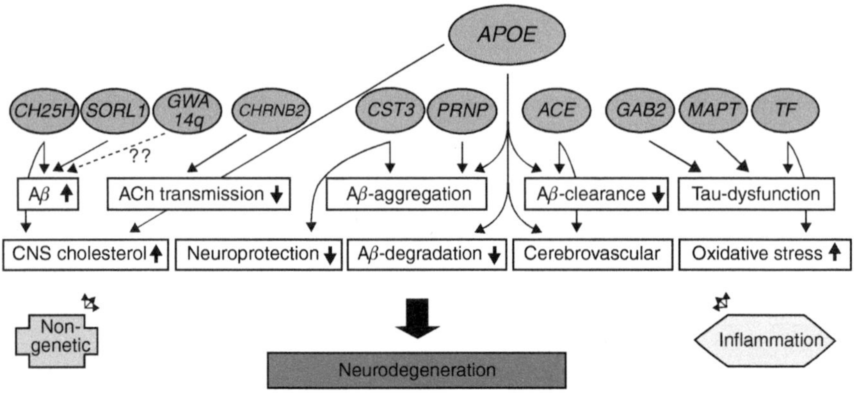

FIG. 1. *Potential pathogenetic roles of AD susceptibility genes.* Simplified summary of the potential pathophysiologic implications of the currently most promising AD susceptibility loci, including those of recent GWAS. Note that none of the genetic associations (except *APOE*) can be considered unequivocally established; the same is true for any of the proposed potential functional implications. Highlighted effects are likely further influenced by currently unknown gene-gene interactions and the contribution of nongenetic factors as well as inflammatory reactions (which themselves are likely genetically controlled). For didactic purposes, not all presumed or established functional implications have been indicated, for example, *GAB2's* role on APP metabolism via binding of Grb2. GWA 14q indicates GWAS locus identified by Bertram *et al.* (2008a). For a more detailed discussion of the potential role of these and other AD candidate genes see Bertram and Tanzi (2008).

popular, but it remains to be seen whether GWAS will live up to these expectations. In the interim, systematic bioinformatics approaches synthesizing the results from both candidate gene and genome-wide analyses will help researchers to keep track of the myriad of genetic association findings to come. One such approach, the AlzGene database, is now available, and already highlights a number of promising AD loci by means of systematic, unbiased meta-analysis.

As should be clear from the above discussion, AlzGene does not aim to deliver the *last* piece in the puzzle that AD genetic epidemiology research is trying to solve. Rather, it attempts to provide a tool that can help researchers of many disciplines decide which piece of the puzzle to try *next.* In the best case scenario, it will also help to sharpen the overall picture of the genetic forces driving AD predisposition and pathogenesis. Eventually, only the concerted efforts of genetics, genomics, proteomics, and clinical disciplines will give rise to new diagnostic and therapeutic targets that, hopefully in the not too distant future, will benefit the millions of patients afflicted with this devastating disorder.

Acknowledgments

Funding for AlzGene was provided by the Cure Alzheimer's Fund (to LB). We are grateful to the Alzheimer Research Forum (http://www.alzforum.org) for hosting AlzGene on their Web site.

References

Allen, N. C., Bagade, S., McQueen, M. B., Ioannidis, J. P., Kavvoura, F. K., Khoury, M. J., Tanzi, R. E., and Bertram, L. (2008). Systematic meta-analyses and field synopsis of genetic association studies in schizophrenia: The SzGene database. *Nat. Genet.* **40,** 827–834.

Bagade, S., Allen, N. C., Tanzi, R. E., and Bertram, L. (2006). PD Gene–A novel and publicly available database of genetic associated studies in Parkinson's disease. 36th Meeting of the Society for Neuroscience (Atlanta, GA), p. 173.

Bergem, A. L., Engedal, K., and Kringlen, E. (1997). The role of heredity in late-onset Alzheimer disease and vascular dementia. A twin study. *Arch. Gen. Psychiatry* **54,** 264–270.

Bertram, L. (2008). Genetic research in schizophrenia: New tools and future perspectives. *Schizophr Bull.* **34,** 806–812.

Bertram, L., Blacker, D., Mullin, K., Keeney, D., Jones, J., Basu, S., Yhu, S., McInnis, M. G., Go, R. C., Vekrellis, K., Selkoe, D. J., Saunders, A. J., *et al.* (2000). Evidence for genetic linkage of Alzheimer's disease to chromosome 10q. *Science* **290,** 2302–2303.

Bertram, L., Lange, C., Mullin, K., Parkinson, M., Hsiao, M., Hogan, M. F., Schjeide, B. M., Hooli, B., Divito, J., Ionita, I., Jiang, H., Laird, N., *et al.* (2008a). Genome-wide association analysis reveals putative Alzheimer's disease susceptibility loci in addition to APOE. *Am. J. Hum. Genet.* **83,** 623–632.

Bertram, L., McQueen, M. B., Mullin, K., Blacker, D., and Tanzi, R. E. (2007). Systematic meta-analyses of Alzheimer disease genetic association studies: The AlzGene database. *Nat. Genet.* **39,** 17–23.

Bertram, L., Schjeide, B. M., Hooli, B., Mullin, K., Hiltunen, M., Soininen, H., Ingelsson, M., Lannfelt, L., Blacker, D., and Tanzi, R. E. (2008b). No association between CALHM1 and Alzheimer's disease risk. *Cell* **135,** 993–994.

Bertram, L., and Tanzi, R. E. (2005). The genetic epidemiology of neurodegenerative disease. *J. Clin. Invest.* **115,** 1449–1457.

Bertram, L., and Tanzi, R. E. (2008). Thirty years of Alzheimer's disease genetics: The implications of systematic meta-analyses. *Nat. Rev. Neurosci.* **9,** 768–778.

Blacker, D., Bertram, L., Saunders, A. J., Moscarillo, T. J., Albert, M. S., Wiener, H., Perry, R. T., Collins, J. S., Harrell, L. E., Go, R. C., Mahoney, A., Beaty, T., *et al.* (2003). Results of a high-resolution genome screen of 437 Alzheimer's disease families. *Hum. Mol. Genet.* **12,** 23–32.

Brewer, G. J. (2007). Iron and copper toxicity in diseases of aging, particularly atherosclerosis and Alzheimer's disease. *Exp. Biol. Med. (Maywood)* **232,** 323–335.

Coon, K. D., Myers, A. J., Craig, D. W., Webster, J. A., Pearson, J. V., Lince, D. H., Zismann, V. L., Beach, T. G., Leung, D., Bryden, L., Halperin, R. F., Marlowe, L., *et al.* (2007). A high-density whole-genome association study reveals that APOE is the major susceptibility gene for sporadic late-onset Alzheimer's disease. *J. Clin. Psychiatry* **68,** 613–618.

Ertekin-Taner, N., Graff-Radford, N., Younkin, L. H., Eckman, C., Baker, M., Adamson, J., Ronald, J., Blangero, J., Hutton, M., and Younkin, S. G. (2000). Linkage of plasma Abeta42 to a quantitative locus on chromosome 10 in late-onset Alzheimer's disease pedigrees. *Science* **290,** 2303–2304.

Gatz, M., Reynolds, C. A., Fratiglioni, L., Johansson, B., Mortimer, J. A., Berg, S., Fiske, A., and Pedersen, N. L. (2006). Role of genes and environments for explaining Alzheimer disease. *Arch. Gen. Psychiatry* **63,** 168–174.

Goate, A., Chartier-Harlin, M. C., Mullan, M., Brown, J., Crawford, F., Fidani, L., Giuffra, L., Haynes, A., Irving, N., James, L., *et al.* (1991). Segregation of a missense mutation in the amyloid precursor protein gene with familial Alzheimer's disease. *Nature* **349,** 704–706.

Grupe, A., Abraham, R., Li, Y., Rowland, C., Hollingworth, P., Morgan, A., Jehu, L., Segurado, R., Stone, D., Schadt, E., Karnoub, M., Nowotny, P., *et al.* (2007). Evidence for novel susceptibility genes for late-onset Alzheimer's disease from a genome-wide association study of putative functional variants. *Hum. Mol. Genet.* **16,** 865–873.

Hamshere, M. L., Holmans, P. A., Avramopoulos, D., Bassett, S. S., Blacker, D., Bertram, L., Wiener, H., Rochberg, N., Tanzi, R. E., Myers, A., Wavrant-De Vrieze, F., Go, R., *et al.* (2007). Genome-wide linkage analysis of 723 affected relative pairs with late-onset Alzheimer's disease. *Hum. Mol. Genet.* **16,** 2703–2712.

Hardy, J., and Selkoe, D. J. (2002). The amyloid hypothesis of Alzheimer's disease: Progress and problems on the road to therapeutics. *Science* **297,** 353–356.

Hoggart, C. J., Clark, T. G., De Iorio, M., Whittaker, J. C., and Balding, D. J. (2008). Genome-wide significance for dense SNP and resequencing data. *Genet. Epidemiol.* **32,** 179–185.

Hu, J., Igarashi, A., Kamata, M., and Nakagawa, H. (2001). Angiotensin-converting enzyme degrades Alzheimer amyloid beta-peptide (A beta); retards A beta aggregation, deposition, fibril formation; and inhibits cytotoxicity. *J. Biol. Chem.* **276,** 47863–47868.

Ioannidis, J. P. (2005). Why most published research findings are false. *PLoS Med.* **2,** e124.

Ioannidis, J. P., Ntzani, E. E., Trikalinos, T. A., and Contopoulos-Ioannidis, D. G. (2001). Replication validity of genetic association studies. *Nat. Genet.* **29,** 306–309.

Kalamida, D., Poulas, K., Avramopoulou, V., Fostieri, E., Lagoumintzis, G., Lazaridis, K., Sideri, A., Zouridakis, M., and Tzartos, S. J. (2007). Muscle and neuronal nicotinic acetylcholine receptors. Structure, function and pathogenicity. *Febs J.* **274,** 3799–3845.

Kauwe, J. S., Wang, J., Mayo, K., Morris, J. C., Fagan, A. M., Holtzman, D. M., and Goate, A. M. (2008). Alzheimer's disease risk variants show association with cerebrospinal fluid amyloid beta. *Neurogenetics.*

Kehoe, P., Wavrant-De Vrieze, F., Crook, R., Wu, W. S., Holmans, P., Fenton, I., Spurlock, G., Norton, N., Williams, H., Williams, N., Lovestone, S., Perez-Tur, J., *et al.* (1999). A full genome scan for late onset Alzheimer's disease. *Hum. Mol. Genet.* **8,** 237–245.

Kennedy, J. L., Farrer, L. A., Andreasen, N. C., Mayeux, R., and St. George-Hyslop, P. (2003). The genetics of adult-onset neuropsychiatric disease: Complexities and conundra? *Science* **302,** 822–826.

Laird, N. M., and Lange, C. (2006). Family-based designs in the age of large-scale gene-association studies. *Nat. Rev. Genet.* **7,** 385–394.

Levy-Lahad, E., Wasco, W., Poorkaj, P., Romano, D. M., Oshima, J., Pettingell, W. H., Yu, C. E., Jondro, P. D., Schmidt, S. D., Wang, K., *et al.* (1995). Candidate gene for the chromosome 1 familial Alzheimer's disease locus. *Science* **269,** 973–977.

Li, H., Wetten, S., Li, L., St. Jean, P. L., Upmanyu, R., Surh, L., Hosford, D., Barnes, M. R., Briley, J. D., Borrie, M., Coletta, N., Delisle, R., *et al.* (2008). Candidate single-nucleotide polymorphisms from a genomewide association study of Alzheimer disease. *Arch. Neurol* **65,** 45–53.

Li, Y. J., Scott, W. K., Hedges, D. J., Zhang, F., Gaskell, P. C., Nance, M. A., Watts, R. L., Hubble, J. P., Koller, W. C., Pahwa, R., Stern, M. B., Hiner, B. C., *et al.* (2002). Age at onset in two common neurodegenerative diseases is genetically controlled. *Am. J. Hum. Genet.* **70,** 985–993.

Liu, Y., and Rohrschneider, L. R. (2002). The gift of Gab. *FEBS Lett.* **515,** 1–7.

Loeffler, D. A., Connor, J. R., Juneau, P. L., Snyder, B. S., Kanaley, L., DeMaggio, A. J., Nguyen, H., Brickman, C. M., and LeWitt, P. A. (1995). Transferrin and iron in normal, Alzheimer's disease, and Parkinson's disease brain regions. *J. Neurochem.* **65,** 710–724.

Lohmueller, K. E., Pearce, C. L., Pike, M., Lander, E. S., and Hirschhorn, J. N. (2003). Meta-analysis of genetic association studies supports a contribution of common variants to susceptibility to common disease. *Nat. Genet.* **33,** 177–182.

McCarthy, M. I., Abecasis, G. R., Cardon, L. R., Goldstein, D. B., Little, J., Ioannidis, J. P., and Hirschhorn, J. N. (2008). Genome-wide association studies for complex traits: Consensus, uncertainty and challenges. *Nat. Rev. Genet.* **9,** 356–369.

Meyer, J. M., and Breitner, J. C. (1998). Multiple threshold model for the onset of Alzheimer's disease in teh NAS-NRC twin panel. *Am. J. Med. Genet.* **81,** 92–97.

Miners, J. S., Ashby, E., Van Helmond, Z., Chalmers, K. A., Palmer, L. E., Love, S., and Kehoe, P. G. (2008). Angiotensin-converting enzyme (ACE) levels and activity in Alzheimer's disease, and relationship of perivascular ACE-1 to cerebral amyloid angiopathy. *Neuropathol. Appl. Neurobiol.* **34,** 181–193.

Myers, A., Holmans, P., Marshall, H., Kwon, J., Meyer, D., Ramic, D., Shears, S., Booth, J., DeVrieze, F. W., Crook, R., Hamshere, M., Abraham, R., *et al.* (2000). Susceptibility locus for Alzheimer's disease on chromosome 10. *Science* **290,** 2304–2305.

Nizzari, M., Venezia, V., Repetto, E., Caorsi, V., Magrassi, R., Gagliani, M. C., Carlo, P., Florio, T., Schettini, G., Tacchetti, C., Russo, T., Diaspro, A., *et al.* (2007). Amyloid precursor protein and Presenilin1 interact with the adaptor GRB2 and modulate ERK 1,2 signaling. *J. Biol. Chem.* **282,** 13833–13844.

Oddo, S., and LaFerla, F. M. (2006). The role of nicotinic acetylcholine receptors in Alzheimer's disease. *J. Physiol. Paris* **99,** 172–179.

Pericak-Vance, M. A., Grubber, J., Bailey, L. R., Hedges, D., West, S., Santoro, L., Kemmerer, B., Hall, J. L., Saunders, A. M., Roses, A. D., Small, G. W., Scott, W. K., *et al.* (2000). Identification of novel genes in late-onset Alzheimer's disease. *Exp. Gerontol* **35,** 1343–1352.

Reiman, E. M., Webster, J. A., Myers, A. J., Hardy, J., Dunckley, T., Zismann, V. L., Joshipura, K. D., Pearson, J. V., Hu-Lince, D., Huentelman, M. J., Craig, D. W., Coon, K. D., *et al.* (2007). GAB2 alleles modify Alzheimer's risk in APOE epsilon4 carriers. *Neuron* **54,** 713–720.

Reynolds, C. H., Garwood, C. J., Wray, S., Price, C., Kellie, S., Perera, T., Zvelebil, M., Yang, A., Sheppard, P. W., Varndell, I. M., Hanger, D. P., and Anderton, B. H. (2008). Phosphorylation regulates tau interactions with SH3 domains of phosphatidylinositol-3-kinase, phospholipase cgamma 1, GRB2 and SRC-family kinases. *J. Biol. Chem.* **283,** 18177–18186.

Rogaev, E. I., Sherrington, R., Rogaeva, E. A., Levesque, G., Ikeda, M., Liang, Y., Chi, H., Lin, C., Holman, K., Tsuda, T., *et al.* (1995). Familial Alzheimer's disease in kindreds with missense mutations in a gene on chromosome 1 related to the Alzheimer's disease type 3 gene. *Nature* **376,** 775–778.

Sarmay, G., Angyal, A., Kertesz, A., Maus, M., and Medgyesi, D. (2006). The multiple function of Grb2 associated binder (Gab) adaptor/scaffolding protein in immune cell signaling. *Immunol. Lett.* **104,** 76–82.

Saunders, A. M., Strittmatter, W. J., Schmechel, D., George-Hyslop, P. H., Pericak-Vance, M. A., Joo, S. H., Rosi, B. L., Gusella, J. F., Crapper-MacLachlan, D. R., Alberts, M. J., *et al.* (1993). Association of apolipoprotein E allele epsilon 4 with late-onset familial and sporadic Alzheimer's disease. *Neurology* **43,** 1467–1472.

Schjeide, B. M., McQueen, M. B., Mullin, K., Divito, J., Hogan, M. F., Parkinson, M., Hooli, B., Lange, C., Blacker, D., Tanzi, R. E., and Bertram, L. (2008). Assessment of Alzheimer's disease case-control associations using family-based methods. *Neurogenetics.*

Schjeide, B.-M. M., Hooli, B., Parkinson, M., Hogan, M. F., Divito, J., Mullin, K., Blacker, D., Tanzi, R., and Bertram, L. (in press). GAB2 as an Alzheimer's disease susceptibility gene: Follow-up of genome-wide association results. *Arch. Neurol.*

Sherrington, R., Rogaev, E. I., Liang, Y., Rogaeva, E. A., Levesque, G., Ikeda, M., Chi, H., Lin, C., Li, G., Holman, K., *et al.* (1995). Cloning of a gene bearing missense mutations in early-onset familial Alzheimer's disease. *Nature* **375,** 754–760.

Smith, M. A., Harris, P. L., Sayre, L. M., and Perry, G. (1997). Iron accumulation in Alzheimer disease is a source of redox-generated free radicals. *Proc. Natl. Acad. Sci. USA* **94,** 9866–9868.

Strittmatter, W. J., Saunders, A. M., Schmechel, D., Pericak-Vance, M., Enghild, J., Salvesen, G. S., and Roses, A. D. (1993). Apolipoprotein E: High-avidity binding to beta-amyloid and increased frequency of type 4 allele in late-onset familial Alzheimer disease. *Proc. Natl. Acad. Sci. USA* **90,** 1977–1981.

Takeda, S., Sato, N., Ogihara, T., and Morishita, R. (2008). The renin-angiotensin system, hypertension and cognitive dysfunction in Alzheimer's disease: New therapeutic potential. *Front. Biosci.* **13,** 2253–2265.

Tanzi, R. E. (1999). A genetic dichotomy model for the inheritance of Alzheimer's disease and common age-related disorders. *J. Clin. Invest.* **104,** 1175–1179.

Tanzi, R. E., and Bertram, L. (2005). Twenty years of the Alzheimer's disease amyloid hypothesis: A genetic perspective. *Cell* **120,** 545–555.

Tohgi, H., Utsugisawa, K., Yoshimura, M., Nagane, Y., and Mihara, M. (1998). Age-related changes in nicotinic acetylcholine receptor subunits alpha4 and beta2 messenger RNA expression in postmortem human frontal cortex and hippocampus. *Neurosci. Lett.* **245,** 139–142.

Wacholder, S., Chanock, S., Garcia-Closas, M., El Ghormli, L., and Rothman, N. (2004). Assessing the probability that a positive report is false: An approach for molecular epidemiology studies. *J. Natl. Cancer Inst.* **96,** 434–442.

Yamamoto, A., Shin, R. W., Hasegawa, K., Naiki, H., Sato, H., Yoshimasu, F., and Kitamoto, T. (2002). Iron (III) induces aggregation of hyperphosphorylated tau and its reduction to iron (II) reverses the aggregation: Implications in the formation of neurofibrillary tangles of Alzheimer's disease. *J. Neurochem.* **82,** 1137–1147.

FRONTOTEMPORAL LOBAR DEGENERATION: INSIGHTS FROM NEUROPSYCHOLOGY AND NEUROIMAGING

Andrea C. Bozoki and Muhammad U. Farooq

Department of Neurology, Michigan State University, East Lansing, Michigan, 48824, USA

Frontotemporal lobar degeneration (FTLD) is a diagnostic term that encompasses three distinctly different syndromes, united by historic as well as pathologic commonalities. We briefly review the origins of the current classification scheme for diagnosing the three major subtypes—frontotemporal dementia, semantic dementia, and primary progressive aphasia, highlighting the differences between subtypes as well as from Alzheimer's disease (AD). We briefly examine current understanding regarding the histopathology and genetics of FTLD but note that there is a poor correspondence of these features with specific subtypes of FTLD. For clinicians and clinical researchers, this implicates the need for other diagnostic strategies. Neuropsychological and neurobehavioral testing currently offers the most sensitive and specific method for identifying subtypes, and discriminating FTLD from other forms of dementia. Multiple studies from the relevant literature

INTERNATIONAL REVIEW OF
NEUROBIOLOGY, VOL. 84
DOI: 10.1016/S0074-7742(09)00410-3

are reviewed, highlighting those findings that are likely to be most valuable to physicians. Finally, we analyze some of the major findings from the large body of work on neuroimaging of FTLD, again focusing on those studies that potentially help discriminate FTLD subtypes or assist with the discrimination of FTLD from AD.

I. Introduction

Frontotemporal dementia used to be considered a rare disorder-one almost never seen by the average clinician. The typical patient with "Pick's disease" (PD) as it was known synonymously was about 65 years old, socially inappropriate, frequently hyperoral and became progressively more demented over a period of 5-7 years before dying.

In 1882, Arnold Pick described the clinical features of a patient with circumscribed lobar atrophy with an aphasic dementia. His initial cases had only gross examination without any histological data. The clinical pattern and its relation to focal atrophy were the basis of the syndrome. In 1911, Alzheimer described the now-characteristic hallmark inclusion bodies, called Pick's bodies. Other histological findings such as ballooned achromatic neurons (Pick cells), gliosis, and superficial cortical spongiform changes were also identified in these patients. These histopathological findings were associated with the circumscribed frontal and anterior temporal atrophy which Pick described in 1882, and the clinical syndrome came to be known as Pick's disease.

We have come a long way from these early findings. We now know that Pick's initial patients belong to a group of related dementias. The preferred term for this rather heterogeneous group is frontotemporal lobar degeneration (FTLD), and it is believed to be the second most common type of dementia in individuals younger than 65 years (Ratnavalli *et al.*, 2002). It remains clinically under-recognized both due to its frequent misdiagnosis as psychiatric disease or Alzheimer's disease (AD), as well as limited recognition of the different entities within the FTLD spectrum. Overall, estimated prevalence ranges from 6 to 12% of dementias (Kertesz, 2006).

Currently, FTLD is divided into three main subgroups; frontotemporal degeneration (FTD), primary progressive aphasia (PPA) (also referred to as progressive nonfluent aphasia; PNFA), and semantic dementia (SD) (Neary *et al.*, 1998). Each subtype of FTLD has distinguishing features and characteristic progression that help to distinguish it from "garden variety" AD. In addition, both neuropsychological testing and neuroimaging have extended the clinician's ability to distinguish FTLD from AD as well as other related dementias.

A. Diagnosis of FTLD

Clinically and pathologically, FTLD is distinct from AD, although the distinction is sometimes clinically challenging (Klatka *et al.*, 1996; Mendez *et al.*, 1993). A proportion of patients who meet the criteria for AD have FTLD confirmed at autopsy, occasionally co-occurring with AD pathology. A recent pathologic study examining the overlap of FTLD and AD clinical phenotypes described coexisting histopathological AD in almost one fourth of FTLD cases (Liscic *et al.*, 2007). Autopsy-proved FTLD cases, alone or in combination with AD, can meet the National Institute of Neurological and Communicative Disorders and Stroke, and the Alzheimer's Disease and Related Disorders Association (NINDS-ADRDA) clinical criteria for AD during life (McKhann *et al.*, 1984; Snowden *et al.*, 2001), suggesting a lack of specificity in the criteria for both diseases. Despite these concerns (or perhaps because of them), several groups have put forward criteria for the clinical diagnosis of FTLD. The sophistication of these classifications and rating scales for FTLD has evolved since the 1980s.

B. Clinical Criteria

Gustafson and Nilsson (1982) studied 57 patients for differential diagnosis between dementias. They used three rating scales for identification of Alzheimer's disease, Pick's disease, and multi-infarct dementia. The rating scale for PD contained nine items, including assessment of early loss of insight, early signs of disinhibition, echolalia, mutism, and amimia.

Lund-Manchester groups (1994) published criteria in 1994 which required the presence of at least two of the following features: loss of personal awareness, strange eating habits, perseveration, and mood change. In addition, patients had to have one or more of the following: frontal executive dysfunction, reduced speech, and preserved visuospatial ability. Finally, the authors of the criteria cited several important supporting features, including onset before age 65, a family history of FTD, early urinary incontinence, motor neuron disease, and (in the late stages) akinesia, rigidity, and tremor.

Neary and colleagues (Neary et al., 1998) incorporated most of the above features and added some others. However, the authors subdivided patients into three separate clinical presentations: frontotemporal dementia (characterized primarily by personality change and disordered social conduct), PNFA (in which patients have difficulty with initiation but not comprehension of speech; see Section II.B), and semantic aphasia (distinguished by impaired understanding of word meaning and/or object identity). The general term FTLD was used to refer to any of the three syndromes.

ECAPD Consortium, 1998. The European Concerted Action on Pick's disease (ECAPD) was based on the review of 50 cases of pathologically verified disease for which adequate clinical, neuropsychological, and neuroimaging data were available. This resulted in the definition of provisional criteria for definite, clinically probable and clinically possible Pick's disease which emphasized the importance of impairment of language and praxic skills.

The Work Group on Frontotemporal Dementia and Pick's Disease (McKhann et al., 2001). An international group of clinical and basic scientists developed these consensus criteria in July 2000. Intended for clinicians rather than researches, these criteria were considerably simpler than the Neary criteria and lumped together all patients under the terminology of FTD, although the authors did note that patients would have either a "behavioral presentation" or a "language presentation."

C. Histopathology

Originally, Pick's disease was classified pathologically as type A (with Pick's bodies), type B (with only swollen neurons), and type C (with only gliosis) (Constantinidis *et al.*, 1974). However, in common usage, Pick's disease referred to patients with atrophy and disinhibition and Pick's bodies on histopathology. It resulted in the impression that Pick's disease was a rare entity and difficult to diagnose before autopsy. Later, it was found that patients with lobar atrophy only rarely have Pick's bodies and often do not show the typical histologic picture on autopsy (Constantinidis *et al.*, 1974). Therefore, the term FTLD came into use and a new label was given to clinical Pick's disease; those cases without the typical histologic picture were called frontal lobe degeneration (FLD) (Gustafson, 1987; Neary *et al.*, 1988). A similar clinical picture was described under the label of "dementia lacking distinctive histology" (DLDH) (Knopman *et al.*, 1990).

There are a few histological features including neuronal loss, gliosis, and superficial linear spongiosis, which are common to all histological subtypes of FTLD. Pick cells occur with variable frequency in all variants. These are called ballooned neurons because of their swollen appearance under light microscopy. These cells show neuronal achromasia and express phosphorylated neurofilaments. Spongiform changes are seen in layers II and III of the cortex. Various distinct features such as Pick's bodies, astrocyte plaques (common in Corticobasal Degeneration), tufted astrocytes (common in Progressive Supranuclear Palsy), and ubiquitin-positive tau-negative inclusions (common in Amyotrophic Lateral Sclerosis) known as motor neuron disease type inclusions or MNDI, have been described but can also occur with each of the FTLD variants and are non-specific. However, MNDI are found in more than half of the FTD cases on autopsy and

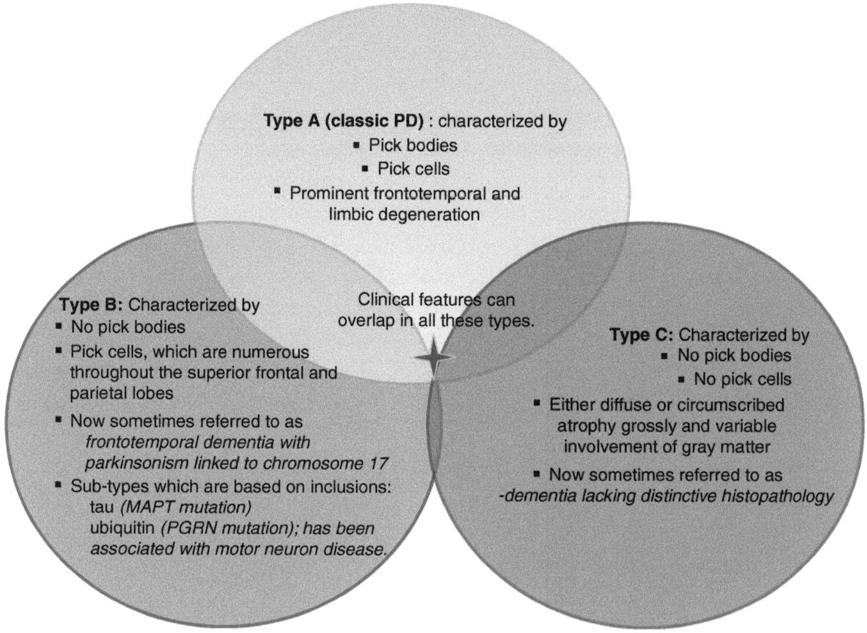

FIG. 1. Classification of Pick's disease based on histopathologic features. It is important to note that all three major clinical phenotypes have been documented with each of the pathologic subtypes.

now form the largest single pathological variety of Pick's complex (Hodges *et al.*, 2004; Kertesz *et al.*, 2005; Munoz *et al.*, 2003) (Fig. 1).

D. GENETICS

FTLD is a "presenile" dementia with a strong genetic component (Chow *et al.*, 1999; Stevens *et al.*, 1998). The first genetic correspondence, a linkage to chromosome 17 q21-22, was discovered in a large family with variable symptomatology of frontotemporal dementia, aphasia, parkinsonism, and amyotrophy (Wilhelmsen *et al.*, 1994). Later, the clinical features and pathology of 12 families resembling sporadic cases were described and the term "dominant inherited chromosomal 17-linked frontotemporal dementia with Parkinsonism" (FTDP-17) was accepted as a definite entity in a consensus conference (Foster *et al.*, 1997). The tau protein was suspected as the candidate gene for mutation and several tau mutations have been identified since then which occur only in familial cases (Hutton *et al.*, 1998). However, tau-positive pathology is not the only substrate of the disease, as the ubiquinated MNDI patients demonstrate (Hodges *et al.*,

2004; Jackson *et al.*, 1996). The phenotypic and pathologic variations of these mutations closely match sporadic disease and provide strong evidence for the cohesion of Pick's complex.

Like histopathology, genetic testing is also not specific for the subtypes of FTLD, with significant syndromic overlap within single mutation types. Tau protein mutations were first reported for the phenotypes of the FTDP-17 complex, which varied widely (some patients with parkinsonism, some with prominent aphasia, some with obvious amyotrophy. Several tau mutations have been discovered so far (Clark *et al.*, 1998; Hutton *et al.*, 1998; Spillantini *et al.*, 1998). Different tau mutations differently alter the biochemical properties of tau isoforms, but so far these mutations do not predict the clinical presentation. In addition, the tau protein is not abnormal in all forms of FTLD. The amount of hyperphosphorylated tau in FTLD can be low and sometimes even absent.

II. Classification of FTLD Subtypes and Their Features

A. Frontotemporal Degeneration

FTD is the most common diagnostic subgroup of FTLD; over half of FTLD patients fall in this category (Hodges *et al.*, 2003). FTD also has the earliest age at onset, often between 55 and 65, and almost never after age 75 (Johnson *et al.*, 2005; Rosso *et al.*, 2003). FTD is characterized by personality and behavioral changes. The initial manifestations can be subtle and difficult to diagnose beginning with apathy and disinterest which is often misdiagnosed as depression (Kertesz, 2003). These patients later develop lack of insight and disinhibition, impulsive and inappropriate behavior, poor financial judgment, changes in *libido* and inappropriate sexual behavior, blunting of appropriate behavioral response and at the extreme, self-destructive behavior (McKhann *et al.*, 2001). Their dietary habits may change, resulting in overeating or eating of only certain types of food. The overall appearance and personal hygiene of these patients is frequently affected as they show little personal concern for their actions. These patients can have problems with their memory, and formal memory testing can be abnormal, but they can track day-to-day events and do not have a typical amnestic syndrome (McKhann *et al.*, 2001). Frontal lobe functions are impaired in these patients but some can continue to perform well on formal measures of frontal tasks early in the disease course. Orientation and episodic memory is also relatively preserved; problems of forgetfulness are mainly due to inattention and poor organization of incoming information (Kertesz, 2003). Language is not affected early in the course of this disease; however, most patients eventually develop decreased verbal output and often progress to mutism in later stages (Kertesz, 2003).

B. Primary Progressive Aphasia or Progressive Nonfluent Aphasia

The term PPA was first used by Mesulam (1982) to describe a group of six patients who developed a slowly progressive aphasic disorder that "began in the presenium" (before age 65). These patients present with language difficulty characterized by a combination of word finding difficulties, abnormal speech patterns, decreased comprehension, and impaired spelling. In the early part of the disease these patients are able to perform their daily activities and have normal memory and visuospatial function. Mesulam's original description stressed the lack of a generalized dementia, and his criteria eventually included a stipulation that the patient's language difficulties should remain the only feature for at least 2 years before more generalized deficits develop (Weintraub *et al.*, 1990). Language deficits ultimately progress to mutism, however, these patients instead of developing a severe dementia in the advanced stages of their disease may have relatively preserved memory and can function well in the community, further helping to distinguish them from patients with AD (Kertesz, 2003). While the typical PPA patient does not have florid psychiatric features the way many FTD patients do, some can develop the behavioral problems seen in patients with other forms of FTLD in the later part of the disease and the distinction may become difficult in advanced cases, when both mutism and inappropriate behavior are seen.

C. Semantic Dementia

Snowden *et al.* described a distinct, fluent form of PPA that is different from the more common nonfluent type and called this semantic dementia (Snowden *et al.*, 1989). In this condition patients slowly and progressively lose the meaning of words; however, fluency remains intact and they are able to carry on a (somewhat empty) conversation. They have difficulty naming and comprehending even single nouns (Hodges *et al.*, 1992; Kertesz, 2003). These patients differ from the fluent aphasia of AD as they have relatively preserved episodic and autobiographical memory with a rather selective loss of semantic memory. These patients retain information that has immediate relevance to their environment or to their person; however, they lose the meanings of other common things. Some of these patients have visual agnosia (Kertesz, 2003). Patients with SD ultimately become nonfluent and even mute. They also develop behavioral problems in the later part of their disease.

We note that some authors categorize PPA and SD together as the language and aphasic variant of FTLD. In this nosologic schema, FTD represents the frontal variant of FTLD and PPA/SD together represent the temporal/parietal variant (Kertesz, 2003).

In addition, there is controversy regarding the separate existence of a fluent form of PPA distinct from both the more common dysfluent PPA and SD (Adlam *et al.*, 2006). In this schema, PPA is thought to present in both fluent and nonfluent forms. Some patients who meet the criteria for fluent PPA show nonverbal and verbal deficits in conceptual knowledge and other tasks that depend upon such knowledge, provided that the measures used allow for the vital impact of concept and typicality. If more emphasis is placed on the prominent language disorder, these individuals can be classified as fluent PPA; if more emphasis is placed on the multimodal pattern of deficits, these individuals can be seen as having SD.

III. Neuropsychological Assessment

A. Distinguishing FTLD from AD

1. *Comments Regarding the MMSE and FTLD*

Although neuropsychologists typically disdain the Folstein Mini Mental State Exam (MMSE; Folstein *et al.*, 1975) as being far too cursory to be useful, it is commonly used by clinicians as a screening tool for the presence of dementia and therefore worthy of comment in this chapter. The MMSE is heavily weighted toward verbal skills and gives relatively short shrift to executive processing. As such, compared with AD, it may underestimate dementia severity for FTD and overestimate it for PPA/SD. Clinicians are especially cautioned not to consider a normal-range MMSE score in a patient with suspected FTD as indicating the absence of disease. A recent study (Osher *et al.*, 2007) which compared FTD and PPA (but did not include AD), found that MMSE scores over time did correlate with changes on a measure of activities of daily living (the Activities of Daily Living Questionnaire or ADL-Q; Johnson *et al.*, 2004). However, PPA patients showed relatively greater declines, suggesting that the MMSE overestimates dementia severity for PPA patients but that the score may accurately measure functional impairment in FTD (Fig. 2).

2. *FTD*

Attempts to distinguish FTD from AD based only on neuropsychological testing have a long and controversial history. Numerous papers on this topic, spanning several decades, have failed to agree on whether it is possible to do this reliably. One group (Gregory *et al.*, 1997) failed to find differences between patient groups in memory, attention, language, and executive abilities, while another (Pasquier *et al.*, 1995) did not find any difference in letter and category verbal fluency (a frequently used bedside test for evaluating frontal lobe function).

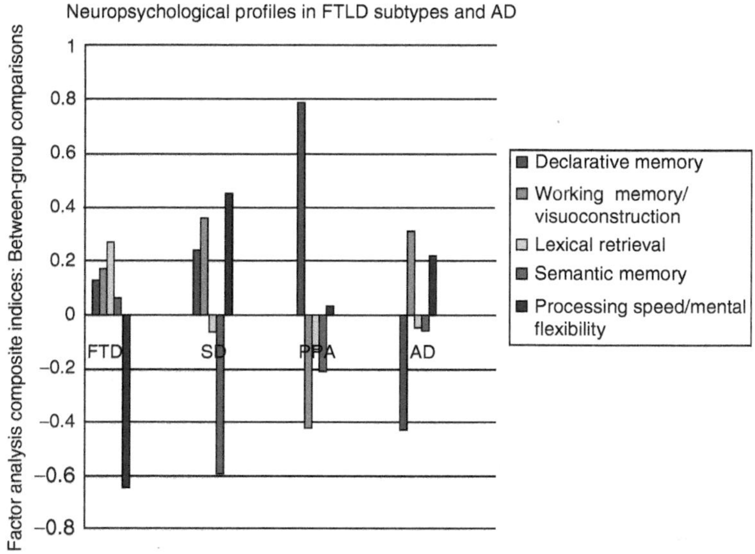

FIG. 2. Factor analysis of a multitest neuropsychological battery demonstrated five major cognitive factors. Between- and within-group analyses of the factors demonstrate that each patient group is associated with a specific profile of neuropsychological impairment, in which each group is significantly worse than the others on at least one factor (data from Libon *et al.*, 2007).

Conversely, several more recent studies have found the opposite; that a double dissociation between AD and FTD exists and can be identified with careful neuropsychological assessment. Perri and colleagues (2005) performed a stepwise discriminant analysis on a number of neuropsychometric variables and determined that subjects with AD were significantly worse on the Rey figure copy (a measure of visuospatial praxis) and significantly better on letter fluency than the FTD group, such that a combination of these two tests plus a measure of apathy (see Section IV) was sufficient to distinguish the two diseases with approximately 83% sensitivity and 82% specificity.

Also in 2005, a group from the UK (Thompson *et al.*, 2005), observing that patients with FTD tend to make different types of mistakes from those with AD, developed a way to assess qualitative performance and error types on their neuropsychometric battery. They found that "FTD patients displayed features associated with frontal lobe dysfunction, such as concrete thought, perseveration, confabulation, and poor organization, which disrupted performance across the range of neuropsychological tests." They advise that keeping track of the *types* and not just the number of errors made in performing neuropsychological tests can greatly enhance discrimination of these patient populations.

Nonetheless, while this chapter finds that FTD and AD can be discriminated neuropsychologically, they agree with the findings of the first two studies cited, that actual scores did not differ significantly between groups, especially on the classic tests of executive functioning. They attribute this to the idea that AD patients are failing at executive tasks "for non-frontal reasons." That is, executive tasks are complex and actually make demands on multiple cognitive domains (particularly memory and visuospatial abilities) that are weak in AD.

This is similar to the findings from the most recent group to weigh in on the matter. This year, a relatively large study (40 FTD, 77 AD, and 91 normal controls) (Giovagnoli *et al.*, 2008) was unable to find a significant difference between the patient groups with regard to executive functions. In this study, the AD patients were significantly more impaired than the FTD group in episodic memory, selective attention, visual perception, visuomotor coordination, and constructive praxis, whereas no differences were found in executive, intellective, and linguistic abilities between the two patient groups. Logistic regression analyses revealed that episodic memory significantly predicted the diagnosis of AD while no executive deficit was able to predict the diagnosis of FTD. The authors hypothesize that this may be due to a more widespread distribution of executive functions in the brain than is currently acknowledged, with other brain regions providing an extended network of functions that are interrelated with prefrontal cortex.

Another reason for the failure of some groups to find differences between FTD and AD may be that not all studies distinguished between FTD and SD in their samples; that is, they included both types of patients in their testing. Because FTD and SD overlap with AD deficits in very different ways, the net effect may have been to "cancel out" the differences between groups. This is the hypothesis of a study which in fact did find a double dissociation between AD, FTD and SD groups in their study (Perry and Hodges, 2000).

3. *PPA*

In contrast to the relatively difficult AD-FTD distinction neuropsychologically, distinguishing between AD and PPA is more straightforward. PPA patients maintain relatively normal episodic memory function—the hallmark decline of early AD, while developing progressively worsening language impairments. The original report on this clinical syndrome was published in 1982 by Mesulam, who reported on six patients with progressively worsening language impairment in the absence of dementia (Mesulam, 1982). Some years later, his lab published a follow-up study (Weintraub *et al.*, 1990) further characterizing three patients with sequential neuropsychological assessments. They demonstrated progressively worsening decline on measures such as the Token Test, sentence repetition and the Boston Naming test. Importantly, cognition was stable over time on measures such as spatial and temporal orientation, design recall, line orientation, and face